Lecture Notes in Mathematics

Edited by A. Dold and B. Eckmann

T0220056

320

Modular Functions of One Variable I

Proceedings International Summer School
University of Antwerp, RUCA
July 17–August 3, 1972

Edited by W. Kuijk

Springer-Verlag
Berlin Heidelberg New York Tokyo

Editor

Willem Kuijk
Rijksuniversitair Centrum Antwerpen, Leerstoel Algebra
Groenenborgerlaan 171, 2020 Antwerpen, Belgium

1st Edition 1973
2nd Corrected Printing 1986

Mathematics Subject Classification (1970): 10D05, 10D25, 10C15, 14K22, 14K25

ISBN 978-3-540-06219-6 Springer-Verlag Berlin Heidelberg New York Tokyo
ISBN 978-0-387-06219-8 Springer-Verlag New York Heidelberg Berlin Tokyo

Printing and binding: Beltz Offsetdruck, Hemsbach/Bergstr.
2146/3140-543210

Preface

An international Summer School on:

"Modular functions of one variable and arithmetical applications"

took place at RUCA, Antwerp University, from July 17 to August 3, 1972.

This book is the first volume (in a series of four) of the Proceedings of the Summer School. It includes the basic course given by A. Ogg, and several other papers with a strong analytic flavour.

Volume 2 contains the courses of R. P. Langlands (ℓ-adic representations) and P. Deligne (modular schemes - representations of GL_2) and papers on related topics.

Volume 3 is devoted to p-adic properties of modular forms and applications to ℓ-adic representations and zeta functions.

Volume 4 collects various material on elliptic curves, including numerical tables.

The School was a NATO Advanced Study Institute, and the organizers want to thank NATO for its major subvention. Further support, in various forms, was received from IBM Belgium, the Coca-Cola Co. of Belgium, Rank Xerox Belgium, the Fort Food Co. of Belgium, and NSF Washington, D.C.. We extend our warmest thanks to all of them, as well as to RUCA and the local staff (not forgetting hostesses and secretaries!) who did such an excellent job.

P. Deligne
W. Kuyk
G. Poitou
J-P. Serre

CONTENTS

SURVEY OF MODULAR FUNCTIONS
OF ONE VARIABLE

by Andrew Ogg

(Notes by F. Van Oystaeyen)

International Summer School on Modular Functions
Antwerp 1972

CONTENTS

INTRODUCTION

In this paper a survey is given of the theory of modular functions of one variable, including Dirichlet series, functional equations, compactifications, Hecke operators, Eisenstein series and the Petersson product. They are notes of an introductory course on the subject at the International Summer School 1972, held at Antwerp University. It is to be noted that much of the material presented is to be found in [7], where complete proofs of most of the theorems occur. However, the material goes beyond [7], and is complementary to it, in the sense that more attention is paid to the relation between Eisenstein series and elliptic curves. Also included is an exposition of some basic features of the work of Artin and Lehner on old and new forms. Little, if no attention at all, is paid to the relation between modular functions and quadratic forms.

The Editor

Ogg-4

1. Elliptic curves and Eisenstein series

Let L be a lattice of $\mathbb{C}$. We can take a basis $\{\omega_1, \omega_2\}$ for L such that $\mathrm{Im}\,(\omega_1/\omega_2) > 0$, i.e. $L = \mathbb{Z}\omega_1 \oplus \mathbb{Z}\omega_2$ and $\tau = \omega_1/\omega_2$ is in the upper half plane H. To a lattice L there is associated an elliptic curve $E = \mathbb{C}|L$ and lattices L, L' define isomorphic elliptic curves E, E' resp., if and only if $L' = tL$ with $t \in \mathbb{C}^*$.

Moreover, $\{\omega_1, \omega_2\}$ and $\{\omega_1', \omega_2'\}$ will define the same lattice L if and only if

$$\omega_1' = a\omega_1 + b\omega_2$$

$$\omega_2' = c\omega_1 + d\omega_2$$

with $\begin{pmatrix} a & b \\ c & d \end{pmatrix} \in SL(2, \mathbb{Z}) = \Gamma'$; indeed, the determinant of the matrix transforming $\{\omega_1, \omega_2\}$ into $\{\omega_1', \omega_2'\}$ cannot be -1 since $\mathrm{Im}(\omega_1/\omega_2)$ and $\mathrm{Im}(\omega_1'/\omega_2')$ are both positive.

The action of Γ' on lattices induces the action of $\Gamma = SL(2,\mathbb{Z})/\{\pm I\}$ in H as follows:

$$\alpha = \begin{pmatrix} a & b \\ c & d \end{pmatrix} \in \Gamma' \qquad \alpha\tau = \frac{a\tau + b}{c\tau + d}.$$

We call Γ <u>the modular group.</u>

In this way we obtain a one-to-one correspondence between the space of isomorphism classes of elliptic curves and $H|\Gamma$. The group Γ' is called the <u>homogeneous modular group</u>. For a subgroup $G' \subset \Gamma'$, of finite index, we define the group G of linear fractional transformations associated with G', by $G = G'/G' \cap \{\pm 1\}$.

The classical Weierstrass form of an elliptic curve is obtained by introducing the $\wp$-functions.

Let L be a lattice of $\mathbb{C}$;

$$\wp(u,L) = \frac{1}{u^2} + \sum_{\omega\in L}' \left(\frac{1}{(u+\omega)^2} - \frac{1}{\omega^2}\right),$$

where $\sum'$ means that the summation is over all $\omega \in L$, $\omega \neq 0$.

For $\omega \in L$, $\wp(u,L) = \wp(u + \omega, L)$. Simple computation shows that,

$$\wp(u,L) = \frac{1}{u^2} + \sum{}' \frac{-2u\omega-u^2}{(u+\omega)^2\omega^2} \qquad \text{and}$$

$$\sum{}' \frac{-2u\omega-u^2}{(u+\omega)^2\omega^2} \ll \sum{}' |\omega|^{-3}, \qquad \text{if}$$

u varies in some compact domain of convergency.

LEMMA 1: <u>The series</u> $\sum'_{n,m\in\mathbb{Z}} |n\tau + m|^{-k}$, $k > 2$, <u>converges uniformly on compact sets</u>.

If we put $k = 3$, the lemma then states that the functions $\wp(u,L)$ converge on compact sets.

For $k = 4,6...$ we define the homogeneous Eisenstein series $G_k(\omega_1, \omega_2) = \sum'_\omega \omega^{-k}$; the inhomogeneous series is derived from it by putting $G_k(\omega_1, \omega_2) = (\frac{2\pi i}{\omega_2})^k\, G_k(\tau)$. Lemma 1 also proves that $G_k(\tau)$ is a holomorphic function in H. We say that $G_k(\omega_1, \omega_2) = G_k(L)$ is a homogeneous function of <u>dimension</u> $-k$, while $G_k(\tau)$ is a function of <u>weight</u> k.

Since $G_k(\omega_1, \omega_2)$ is invariant under the action of Γ' we get:

$$G_k(\frac{a\tau+b}{c\tau+d})(\frac{2\pi i}{c\omega_1+d\omega_2})^k = G_k(\tau)(\frac{2\pi i}{\omega_2})^k,$$

$$G_k(\frac{a\tau+b}{c\tau+d}) = (c\tau+d)^k\, G_k(\tau).$$

Substituting the expansion of $1/(1+\frac{u}{\omega})^2$, in

$$\wp(u,L) = \frac{1}{u^2} + \sum{}'_{\omega\in L} \left[\frac{1}{(1+\frac{u}{\omega})^2} - 1\right] \omega^{-2}, \qquad \text{we get}$$

$$\wp(u,L) = \frac{1}{u^2} + \sum{}' \omega^{-2}[-2(\tfrac{u}{\omega}) + 3(\tfrac{u}{\omega})^2 - 4(\tfrac{u}{\omega})^3 + ..].$$

The odd-power terms cancel since the summation is over $\omega \in L$, $\omega \neq 0$, so we find

$$\wp(u,L) = \frac{1}{u^2} + 3G_4u^2 + 5G_6u^4 + 7G_8u^6 + \ldots$$

$$\wp'(u,L) = \frac{-2}{u^3} + 6G_4u + 20G_6u^3 + \ldots$$

$$\wp'(u,L)^2 = \frac{4}{u^6} - 24\frac{G_4}{u^2} - 80G_6 + \ldots$$

$$\wp(u,L)^3 = \frac{1}{u^6} + 9\frac{G_4}{u^2} + 15G_6 + \ldots$$

Hence, $\wp'(u,L)^2 - 4\wp(u,L)^3 = -60\frac{G_4}{u^2} - 140G_6 + \ldots$. Now

$\wp'(u,L)^2 - 4\wp(u,L)^3 = -g_2\wp(u,L) - g_3 +$ rest, where $g_2 = 60G_4$, $g_3 = 140G_6$ and the "rest" is a double periodic function with no poles and a zero at the origin, so it has to be zero, and the equation for the $\wp$-function:

$$\wp'(u)^2 = 4\wp(u)^3 - g_2\wp(u) - g_3, \quad \text{follows.}$$

PROPOSITION 1: $\wp'(u,L)^2 = 4(\wp(u,L) - \rho_1)(\wp(u,L) - \rho_2)(\wp(u,L) - \rho_3)$ _with_ ρ_1, ρ_2, ρ_3 _distinct_.

PROOF. Since $\wp'(u,L)$ is an odd function it has zeroes in the points of order two on the elliptic curve, $\rho_1 = \wp(\omega_1/2), \rho_2 = \wp(\omega_2/2), \rho_3 = \wp(\omega_1+\omega_2/2)$, and obviously there cannot exist other zeroes.

COROLLARY. _The discriminant_

$\Delta(\omega_1,\omega_2) = 16(\rho_1-\rho_2)^2\,(\rho_2-\rho_3)^2\,(\rho_3-\rho_1)^2 = g_2^3 - 27g_3^2$, _is never zero_.

Since Δ has no zeroes in $\mathbb{C}$ we can define $j(\omega_1,\omega_2) = (12g_2)^3/\Delta$. The function j only depends on the class of elliptic curves containing $E = \mathbb{C}|L$, i.e. isomorphic curves have the same value for j. So we write $j(\omega_1,\omega_2) = j(E)$ and we call this function the modular invariant of the elliptic curve.

More generally, for a function $F(\omega_1,\omega_2)$ of dimension -k we write

$F(\omega_1,\omega_2) = (\frac{2\pi i}{\omega_2})^k f(\tau)$, and then $f(\tau)$ is a function of weight k. For example, $G_k(\tau)$, with $k = 4,6,\ldots$ has weight k

$\Delta(\tau)$ has weight 12

$j(\tau)$ has weight 0.

These functions are holomorphic in H, and they have period 1, $f(\tau) = f(\tau+1)$. In general, if $f(\tau)$ is a function with period 1, we may write $\sum_{n=-\infty}^{+\infty} a_n q^n$, $q = e^{2\pi i\tau}$ as its Fourier series. We say that $f(\tau)$ is meromorphic at infinity if $a_n = 0$ for $n \leq -N$; $f(\tau)$ is said to be holomorphic at infinity if $a_n = 0$ for $n < 0$, and $f(\tau)$ vanishes at ∞ when $a_n = 0$, $n \leq 0$. The examples G_k, Δ, are holomorphic near ∞. Now, we look at the expansions of G_k, Δ, j. For $k = 4,6,\ldots$ we have

$$G_k(\tau) = (2\pi i)^{-k} \sum'_{n,m\in\mathbb{Z}} (n\tau+m)^{-k}$$

and for k = 2, we put

$$G_2(\tau) = (2\pi i)^{-2} \sum_n \sum'_m (n\tau+m)^{-2}.$$

Hence,

$$G_k(\tau) = (2\pi i)^{-k} \sum_n \sum'_m (n\tau+m)^{-k}, \quad k = 2,4,\ldots$$

$$= \frac{2\zeta(k)}{(2\pi i)^k} + 2(2\pi i)^{-k} \sum_{n=1}^{\infty} \sum_{m\in\mathbb{Z}} (n\tau+m)^{-k}.$$

Now, $\sum_m (\tau+m)^{-2}$ is periodic with period 1 and has a double pole at 0,

$$\sum_m (\tau+m)^{-2} = \frac{\pi^2}{\sin^2\pi\tau} = \pi^2/(\frac{e^{\pi i\tau} - e^{-\pi i\tau}}{2i})^2 = \frac{q(2\pi i)^2}{(1-q)^2} = (2\pi i)^2 \sum_{m=1}^{\infty} mq^m.$$

By repeated derivation we get

$$-2\sum_m (\tau+m)^{-3} = (2\pi i)^3 \sum_{m=1}^{\infty} m^2 q^{m-1}$$

$$\cdots\cdots\cdots\cdots\cdots\cdots$$

$$(-1)^k (k-1)! \sum_m (\tau+m)^{-k} = (2\pi i)^k \sum_{m=1}^{\infty} m^{k-1} q^m$$

using $dq|d\tau = 2\pi iq$.

Substituting this in $G_k(\tau)$, taking into account that $(n\tau+m)^{-k} = n^{-k}(\tau+\frac{m}{n})^{-k}$ we get, $G_k(\tau) = \frac{2\zeta(k)}{(2\pi i)^k} + \frac{2}{(k-1)!} \sum_{m=1}^{\infty} m^{k-1} q^{nm}$,

rewriting this with $nm = \nu$ we find

Ogg-8

$\sum_{n=1}^{\infty} \sigma_{k-1}(n)q^n = \sum_{n=1}^{\infty} q^n \sum_{d|n} d^{k-1} = \sum_{n=1}^{\infty} \frac{n^{k-1}q^n}{1-q^n}$,

so, $G_k(\tau) = \frac{2\zeta(k)}{(2\pi i)^k} + \frac{2}{(k-1)!}\sum_{n=1}^{\infty} \sigma_{k-1}(n)q^n$ is holomorphic at infinity,

hence Δ is too. The Eisenstein series can be written as

$$G_k(\tau) = -\frac{B_k}{k!} + \frac{2}{(k-1)!}\sum_{n=1}^{\infty} \sigma_{k-1}(n)q^n,$$

where B_k is the k-th Bernouilli-number $B_2 = 1/6$, $B_4 = -1/30$, $B_6 = 1/42$ ...; they are defined by $\frac{x}{2}\operatorname{cotg}\frac{x}{2} = 1 + \sum_{n=1}^{\infty} (-1)^n \frac{B_{2n}x^{2n}}{(2n)!}$. Usually the Eisenstein series are normalized by multiplying it with a constant, such that the series equals 1 for $q = 0$, we get the normalized series, $E_k(\tau) = \frac{-k!}{B_k} G_k(\tau) = 1 - \frac{2k}{B_k}\sum_{n=1}^{\infty} \sigma_{k-1}(n)q^n$, and for k=2,

$E_2(\tau) = 1 - 24\sum_{n=1}^{\infty} \sigma_1(n)q^n$. Since, $E_4 = 30.4!\cdot G_4 = 12g_2$, and $E_6 = -42.6!\cdot G_6 = 216g_3$, we have

$$\Delta(\tau) = g_2^3 - 27g_3^2 = (E_4^3(\tau) - E_6^2(\tau))/(12)^3$$

$$= \frac{(1+240\sum \sigma_3(n)q^n)^3 - (1-504\sum \sigma_5(n)q^n)^2}{1728}.$$

We see that the constant term is zero, thus $\Delta(\tau)$ vanishes at infinity, $\Delta(\tau) = q + \sum_{n=2}^{\infty} \tau(n)q^n$, with $\tau(n) \in \mathbb{Z}$. Moreover, we have the following product formula,

$$\Delta(\tau) = q \prod_{n=1}^{\infty} (1-q^n)^{24},$$

(this will be proved later). For the modular invariant j, we derive that

$$j(\tau) = \frac{E_4(\tau)^3}{\Delta(\tau)} = \frac{1}{q} + 744 + \sum_{n=1}^{\infty} c(n)q^n$$

with $c(n) \in \mathbb{Z}$.

REMARK: Let $\alpha = \begin{pmatrix} a & b \\ c & d \end{pmatrix} \in \Gamma'$, and $\tau' = \frac{a\tau+b}{c\tau+d}$. Put $\tau = x+iy$, $\tau' = x'+iy'$. Then $d\tau' = d\tau/(c\tau+d)^2$ and $y' = y/|c\tau+d|^2$. It follows that $f(\tau)$ has weight 2 if and only if $f(\tau)\,d\tau$ is Γ-invariant. Further

$\frac{i}{2}\frac{d\tau \wedge \overline{d\tau}}{y^2} = \frac{dx \wedge dy}{y^2}$, is invariant under $GL^+(2, \mathbb{R})$ and this defines an invariant measure on H.

2. Fundamental domain, compactification etc. ...

Let k be a fixed integer, $\alpha = \begin{pmatrix} a & b \\ c & d \end{pmatrix} \in GL^+(2, \mathbb{R})$, we define an operation $f|_k\alpha$, of α on functions $f(\tau)$, $\tau \in H$, as follows:

$$f|_k\alpha\ (\tau) = (ad-bc)^{k/2}\ (c\tau+d)^{-k}\ f(\frac{a\tau+b}{c\tau+d}).$$

Note that diagonal matrices act trivially;
$f|_k \begin{pmatrix} a & 0 \\ 0 & a \end{pmatrix} (\tau) = f(\tau)$, $a > 0$. Passing to the corresponding homogeneous function $F(\omega_1,\omega_2) = (2\pi i)^k\ \omega_2^{-k}\ f(\omega_1/\omega_2)$ we get

$$(F \circ \alpha) \begin{pmatrix} \omega_1 \\ \omega_2 \end{pmatrix} = (2\pi i)^k\ (c\omega_1+d\omega_2)^{-k}\ f(\frac{a\omega_1+b\omega_2}{c\omega_1+d\omega_2})$$

$$= (2\pi i)^k\ (ad-bc)^{-k/2}\ \omega_2^{-k}\ f|_k\alpha\ (\tau).$$

Thus, the correspondence $f \leftrightarrow F$ induces $f|_k\alpha \leftrightarrow |\alpha|\ F \circ \alpha$, and it follows that $(f|_k\alpha)\ |_k\beta = f\ |_k\alpha\beta$. Let G be a subgroup of $GL^+(2, \mathbb{R})$, f will be a <u>modular form of weight k</u> if

1° $f|_k\alpha = f$ for all $\alpha \in G$,

2° certain regularity conditions are forfilled.

Obviously, it will suffice to check $f|_k\alpha = f$ for a set of generators of the group G. Now, if $G = \Gamma$, the modular group, then the condition 2° is as follows:

$f(\tau)$ is holomorphic in H and at ∞,

$f(\tau) = \sum_{n=0}^{\infty} a_n q^n$, $q = e^{2\pi i\tau}$.

If $a_0 = 0$, then the form is said to be a cusp form, i.e. it vanishes at ∞. A <u>modular function</u> for Γ will simply be a function of weight 0 (invariant) for Γ and meromorphic in H and at ∞. If f, g are modular forms of weight k_1, k_2 resp., then fg has weight $k_1 + k_2$. If f, g are modular forms of the same weight then $f(\tau)/g(\tau)$ is a modular function. Applying this to our former examples we get that $G_k(\tau)$ is a modular form of weight k for Γ, $\Delta(\tau)$ is a cusp-form of weight 12 and $j(\tau)$ is a modular function.

We wish to give H/Γ an analytic structure such that the projection, $H \to H/\Gamma$, is holomorphic. This will lead us to the fundamental domain of the group Γ. Put $T(\tau) = \tau+1$, $S(\tau) = -1/\tau = \tau'$, so $T = \begin{pmatrix} 1 & 1 \\ 0 & 1 \end{pmatrix}$, $S = \begin{pmatrix} 0 & -1 \\ 1 & 0 \end{pmatrix} \in SL(2, \mathbb{Z})$. Let D be the domain: $\{\tau \in \mathbb{C} \mid |\tau| > 1, |R(\tau)| < 1/2\}$. If $\tau \notin D$, then there exists an m such that $T^m(\tau)$ is in D, or $T^m(\tau)$ is such that $|T^m(\tau)| < 1$. Assume then $\tau \notin D$, $|\tau| < 1$, $|R(\tau)| < 1/2$. In this case $y' = y|\,|\tau|^2$ is higher than y and, repeating the procedure we get $\{\tau, \tau_1, \tau_2 \ldots\}$ a set of G-equivalent ($G = \langle S, T \rangle$) points with $y < y_1 < y_2 < \ldots$.

LEMMA 2. <u>A sequence of equivalent points, each "higher" than their predecessor, is finite</u>.

PROOF: Let $\tau' = \frac{a\tau+b}{c\tau+d}$, $y' = y|\,|c\tau+d|^2$ and $y' > y$. Then,

$$1 > |c\tau+d|^2 = (cx+d)^2 + c^2y^2,$$

but for given x, y, there are only finitely many choices for c, d possible, which proves the lemma.

Now, take D to be the domain,

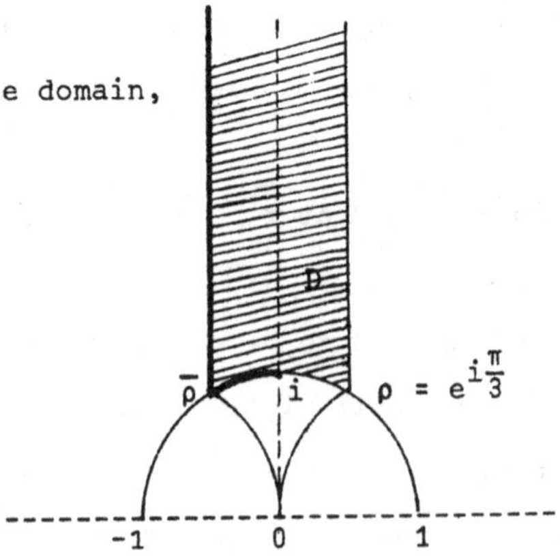

with only the boundary left of the y-ax in it.

LEMMA 3. *If* $\tau' = \frac{a\tau+b}{c\tau+d}$, $\alpha = \begin{pmatrix} a & b \\ c & d \end{pmatrix} \in \Gamma'$, *then* $\tau, \tau' \in D$ *implies* $\tau = \tau'$.

Conclusions: 1. Γ is generated by T and S.

2. D is a fundamental domain for Γ, $H = \cup_{\gamma \in \Gamma} \gamma D$, disjoint.

3. The only fixed points under Γ in D, are the corners of the fundamental domain, i.e., i, $\bar{\rho}$, with stability groups of order 2, 3 resp.

We are going to make a Riemann surface out of H/Γ such that $H \to H/\Gamma$ is holomorphic. The projection $H \to H/\Gamma$ is locally one-one except at points Γ-equivalent to i, ρ. Near the good points we take $z = \tau$ for a local parameter.

Near i we take $z = (\frac{\tau - i}{\tau + i})^2$ and near $\bar{\rho}$ we take $z = (\frac{\tau-\bar{\rho}}{\tau-\rho})^3$. Choosing $q = e^{2\pi i\tau}$ as a local parameter at ∞ on $X = H/\Gamma \cup \{\infty\}$, we get a compact Riemann surface of genus 0.

A modular function is then a meromorphic function on $X = \widehat{H/\Gamma}$. The modular invariant,

$$j(\tau) = \frac{1}{q} + 744 + \sum_{n=1}^{\infty} c(n)q^n \quad (c(n) \in \mathbb{Z})$$

has just one simple pole at ∞ and j may be considered to be a one-to-one map $H/\Gamma \to C$.

If f, g are modular forms of the same weight k, then f/g is a meromorphic function on X and, as such, it has as many zeroes as poles on X which means that f and g have exactly the same number of zeroes in D, (when measured for the adjusted local variables). For example, Δ has weight 12 and has only a simple zero at ∞. Then, for any form f of weight k, Δ^k/f^{12} has a k-multiple zero at ∞ and therefore f has k/12 zeroes in D, so,

$$\bullet \quad k/12 = \frac{1}{3}\,\mathrm{ord}_{\bar{\rho}}(f) + \frac{1}{2}\,\mathrm{ord}_i(f) + \sum_{i,\bar{\rho} \neq \tau \in D} \mathrm{ord}_\tau(f).$$

Let $M(\Gamma,k)$ denote the space of modular forms of weight k for Γ; denote $S(\Gamma,k)$ for the cusp forms. From $\bullet$ it follows directly that,

$$M(\Gamma, k) = 0 \quad \text{for} \quad k < 0 \quad \text{and} \quad k = 2,$$

$$M(\Gamma, 0) = \mathbb{C} \quad \text{and} \quad \dim M(\Gamma, k) = 1 \quad \text{for} \quad k = 4, 6, 8, 10,$$

hence $M(\Gamma, k) = \mathbb{C}\, E_k$ in these cases. In general, if $f \in S(\Gamma, k)$, then f/Δ is an element of $M(\Gamma, k-12)$. It is easily seen that $M(\Gamma, k-12) \cong S(\Gamma, k)$, the isomorphism being given by multiplication with Δ. We get an exact sequence (for $k \geq 4$)

$$0 \to S(k) \to M(k) \to \mathbb{C} \to 0$$

and this proves the dimension formula.

PROPOSITION 2: *If* $k > 0$, *then we have*

$$\dim M(\Gamma,k) = 1 + [k/12] \quad \textit{if} \quad k \not\equiv 2 \bmod 12$$

$$\dim M(\Gamma,k) = [k/12] \quad \textit{if} \quad k \equiv 2 \bmod 12.$$

COROLLARY: *The above proposition yields relations* $E_8 = E_4^2$, $E_{10} = E_4 E_6$, *and, using the expansions of these functions we get identities for their Fourier coefficients.*

PROPOSITION 3: *Let* $f \in M(\Gamma,k)$, *then* f *is an isobaric polynomial in* E_4, E_6, i.e.

$$f(\tau) = \sum_{k = 4a+6b} c_{a,b}\, E_4^a(\tau)\, E_6^b(\tau).$$

3. Dirichlet series and functional equations

Take $\lambda, k > 0$. Let there be given sequences a_n, b_n with $a_n, b_n = O(n^c)$. Put:

$$f(\tau) = \sum_{n=0}^{\infty} a_n\, e^{2\pi i n\tau/\lambda}$$

$$g(\tau) = \sum_{n=0}^{\infty} b_n\, e^{2\pi i n\tau/\lambda}$$

$$\phi(s) = \sum_{n=1}^{\infty} a_n\, n^{-s}$$

$$\psi(s) = \sum_{n=1}^{\infty} b_n\, n^{-s}$$

$$\Phi(s) = \left(\frac{2\pi}{\lambda}\right)^{-s} \Gamma(s)\, \phi(s)$$

$$\Psi(s) = \left(\frac{2\pi}{\lambda}\right)^{-s} \Gamma(s)\, \psi(s).$$

THEOREM 1: <u>The following two statements are equivalent:</u>

(A) $\Phi(s) + \frac{a_0}{s} + \frac{b_0}{k-s}$ <u>is entire and bounded in every vertical strip within the upper half plane, and we have the functional equation</u> $\Phi(k-s) = \Psi(s)$.

(B) $f(-1/\tau) = \left(\frac{\tau}{i}\right)^k g(\tau)$.

PROOF: $$\Phi(s) = \sum_{n=1}^{\infty} a_n \left(\frac{2\pi n}{\lambda}\right)^{-s} \int_0^{\infty} t^s e^{-t} \frac{dt}{t}$$

$$= \sum_{n=1}^{\infty} a_n \int_0^{\infty} e^{-\frac{2\pi n t}{\lambda}} t^s \frac{dt}{t}, \quad \text{and}$$

because of the absolute convergence for $\mathcal{R}(s)$ large,

$$= \int_0^{1} t^s (f(it) - a_0) \frac{dt}{t}.$$

This integral is improper at both ends, but since $f(it) - a_0 = O(e^{-ct})$ as $t \to \infty$, for some $c > 0$, we see that $\int_1^{\infty} t^s (f(it) - a_0) \frac{dt}{t}$ converges uniformly on vertical strips and hence is E.B.V. We first prove (B) $\Rightarrow$ (A)

$$\int_0^{1} t^s (f(it) - a_0) \frac{dt}{t} = -\frac{a_0}{s} + \int_1^{\infty} t^{1-s} f(i/t) \frac{dt}{t^2}$$

(by transforming $t \to 1/t$)

$$= -\frac{a_0}{s} + \int_1^{\infty} t^{k-s} (g(it) - b_0) \frac{dt}{t} - \frac{b_0}{k-s}.$$

Thus, $\Phi(s) + \frac{a_0}{s} + \frac{b_0}{k-s} = \int_1^{\infty} [t^s (f(it) - a_0) + t^{k-s} (g(it) - b_0)] \frac{dt}{t}$ is EBV, and $\Phi(s) = \Psi(k-s)$ which yields (A).

Next assume (A). By Mellin inversion,

$$f(it) - a_0 = \frac{1}{2\pi i} \int_{\sigma = \mathcal{R}(s) = c > 0} t^{-s} \Phi(s)\, ds,$$

where c is large enough, such that s is in the domain of absolute convergence of $\phi(s)$. We shift the line of integration to the left, past 0, and in doing so, we pick up residues $Cb_0\ t^{-k}$ at $s = k$ and $-a_0$ at $s = 0$. Then,

$$f(it) - Cb_0\ t^{-k} = \frac{1}{2\pi i} \int_{\sigma=d\ <\ 0} t^{-s}\ \Phi(s)\ ds$$

$$= \frac{1}{2\pi i} \int_{\sigma=d\ <\ 0} t^{-s}\ \Psi(k-s)\ ds$$

$$= \frac{1}{2\pi i} \int_{\sigma=c'\ >\ k} t^{-(k-s)}\ \Psi(s)\ ds \qquad (s \to k-s)$$

$$= t^{-k}\ (g(i/t) - b_0),$$

and $f(it) = t^{-k}\ g(i/t)$.

For example, consider the Θ-series:

PROPOSITION 4: $\Theta(\tau) = \frac{1}{2}\sum_{n=-\infty}^{+\infty} e^{\pi i\tau n^2}$, <u>has functional equation</u> $\Theta(-1|\tau) = (\frac{\tau}{i})^{1/2}\ \Theta(\tau)$, i.e., $\lambda = 2$, $k = \frac{1}{2}$ <u>in this case</u>.

We have $\Theta(\tau) = \frac{1}{2} + \sum_{n=1}^{\infty} e^{\pi i n^2 \tau}$, where the sum corresponds with $\zeta(2s)$. Put $Z(s) = \pi^{-s}\ \Gamma(s)\ \zeta(2s)$; then we get a functional equation $Z(s) = \pi^{-s}\ \Gamma(s/2)\ \zeta(s) = Z(1-s)$ and $Z(s) + \frac{1}{s} + \frac{1}{1-s}$ is E.B.V. We see that $\zeta(s)$ is determined by its functional equation since $\Theta(\tau)$ is determined by its functional equation.

Take

$$G_k(\tau) = \frac{2\zeta(k)}{(2\pi i)^k} + \frac{2}{\Gamma(k)} \sum_{k=1}^{\infty} \sigma_{k-1}\ (n)\ q^n,$$

we study the functional equation in the cases $k = 2$, $k = 4, 6\ldots$.

1. $k = 4, 6\ldots$

$$f_k(\tau) = -\frac{B_k}{2k}\ E_k(\tau) = -\frac{B_k}{2k} + \sum_{n=1}^{\infty} \sum_{m=1}^{\infty} m^{k-1}\ q^{mn}.$$

The sum is associated with the Dirichlet series

$$\sum_{n=1}^{\infty} \sum_{m=1}^{\infty} m^{k-1}\ (nm)^{-s} = \zeta(s)\ \zeta(s+1-k).$$

Put
$$\Phi_k(s) = (2\pi)^{-s}\,\Gamma(s)\,\zeta(s)\,\zeta(s+1-k)$$
$$= \frac{\zeta(1-s)\,\zeta(s+1-k)}{2\cos\frac{s\pi}{2}}$$
$$\left(\text{since } (2\pi)^{-s}\,\Gamma(s)\,\zeta(s) = \zeta(1-s)/2\cos\tfrac{s\pi}{2}\right)$$
$$= (-1)^{k/2}\,\Phi_k\,(k-s).$$

So we get a functional equation for $G_k(\tau)$, using the equation for the ζ-function. Now since $k \geq 4$, we see that $\zeta(s)$ has a pole at $s = 1$ and $\zeta(s+1-k)$ has a pole at $s = k$, but as $\zeta(k) = 0$ this pole is cancelled out. In this way we proved the functional equation for E_k,
$$E_k\,(-1/\tau) = \left(\frac{\tau}{i}\right)^k (-1)^{k/2}\,E_k(\tau) = \tau^k\,E_k(\tau).$$

2. The case $k = 2$ (Chowla - Weil). Here,
$$\Phi_2(s) = (2\pi)^{-s}\,\Gamma(s)\,\zeta(s)\,\zeta(s-1)$$
$$= -\Phi_2\,(2-s)$$

is such that, $\zeta(s)$ has a pole at $s = 1$ and $\zeta(s-1)$ has a pole at $s = 2$, and hence this pole is not cancelled out. So we get more residues of $\Phi_2(s)$ to take into account in this case;

$$s = 2 : (2\pi)^{-2}\,\frac{\pi^2}{6} = \frac{1}{24} = \text{res}(s = 2) = \text{res}(s = 0),$$

the last equality follows from the first by the functional equation.

$$s = 1 : (2\pi)^{-1}\,\zeta(0) = -\frac{1}{4}\,\pi\,.$$

Thus,
$$f_2(\tau) = -\frac{1}{24} + \frac{1}{2\pi i}\int_{\sigma=c>2} \left(\frac{\tau}{i}\right)^{-s} \Phi_2(s)\,ds$$

and repeating an analogous proof as before we get

$$f_2(\tau) = -1/24 + \text{res}(s=2) + \text{res}(s=1) + \text{res}(s=0)$$
$$+ \frac{1}{2\pi i}\int_{\sigma=d<0} \left(\frac{\tau}{i}\right)^{s-2} (-\Phi_2\,(s))\,ds$$

$$f_2(\tau) = -\frac{\tau^{-2}}{24} + \frac{\tau^{-1}}{4\pi i} + \tau^{-2} \times \frac{1}{2\pi i} \int (\tfrac{\tau}{i})^{s} \Phi_2(s)\, ds$$

$$= \frac{1}{4\pi i\tau} + \tau^{-2} f_2\ (-1/\tau).$$

The functional equation for f_2 is then,

$$f_2\ (-1/\tau) = \tau^2 f_2(\tau) - \tau/4\pi i$$

and, with $f_2 = -\frac{B_2}{4} E_2 = -\frac{1}{24} E_2$, we have

$$E_2(-1/\tau) = \tau^2 E_2(\tau) + \frac{12\tau}{2\pi i}.$$

From the foregoing we can derive a functional equation for Dedekind's function:

$$\eta(\tau) = e^{2\pi i\tau/24}\ \Pi_{n=1}^{\infty}\ (1 - q^n).$$

Logarithmic derivation of $\eta(\tau)$ gives

$$\frac{d}{d\tau} \log \eta(\tau) = \frac{2\pi i}{24} - \sum_{n=1}^{\infty} \frac{nq^n}{1-q^n}\, 2\pi i$$

$$= \frac{2\pi i}{24} E_2(\tau) = -2\pi i\ f_2(\tau).$$

Hence, the functional equation for f_2 yields a transformation formula for $\log \eta(\tau)$:

$$\log \eta(-1/\tau) = \log \eta(\tau) + \frac{1}{2} \log \frac{\tau}{i} + C,$$

evaluating in $\tau = i$ we get $C = 0$. (The log x-function should be well defined, i.e., real for real x).

Thus,

$$\eta(-1/\tau) = \sqrt{\frac{\tau}{i}}\ \eta(\tau)$$

$$\eta(\tau+1) = e^{2\pi i/24}\ \eta(\tau),$$

and it follows that $\eta^{24}(\tau)$ is a cusp form of weight 12 for Γ, therefore $\eta^{24}\ (\tau) = \alpha\Delta$, $\alpha \in \mathbb{C}$. Since the coefficients of q in both series are equal, $\alpha = 1$. We proved, $\Delta = q\ \Pi_{n=1}^{\infty}\ (1-q^n)^{24}$.

APPLICATION: Let $f(\tau)$ be a modular function of weight k, then,

$$f'(\tau) - \frac{2\pi ik}{12} f(\tau)\, E_2(\tau),$$

is a modular form of weight $k + 2$.

4. Hecke operators

Let F be an homogeneous function of lattices of dimension $-k$, i.e., $F(tL) = t^{-k} F(L)$. For $n = 1, 2, \ldots$, we define operators $T'(n)$,

$$F \circ T'(n)\ (L) = \sum_{[L:L']=n} F(L');$$

$F \circ T'(n)$ is still homogeneous and of dimension $-k$.

THEOREM 2: *The operators* $T'(n)$ *and* $T'(m)$ *commute, and satisfy*

$$T'(n)\ T'(m) = \sum_{d|n,m} d^{1-k}\ T'(\frac{nm}{d^2}).$$

PROOF: 1. If $(n,m) = 1$, then $T(nm)$ is a summation over sublattices $L'' \subset L$ of index nm and each of these is contained in a unique sublattice $L' \subset L$ with $[L:L'] = n$, so then $T'(n)\ T'(m) = T'(nm) = T'(m)\ T'(n)$.

2. $n = p$, $m = p^r$. Let L'' be a lattice of index p^{r+1} in L. If $L'' \subset L_1'$ and $L'' \subset L_2'$, where L_1', L_2' are lattices of index p then

$$L'' \subset L_1' \cap L_2' = pL,$$

and in this case L'' occurs $p+1$ times in the sum, $L'' = pL'''$ with $[L : L'''] = p^{r-1}$. Therefore

$$F \circ T'(p) \circ T'(p^r)\ (L) = \sum_{[L:L'']=p^{r+1}} F(L'') + p\sum_{[L:L''']=p^{r-1}} F(pL''')$$

$$= F \circ T'\ (p^{r+1}) + p^{1-k}\ F \circ T'\ (p^{r-1})$$

or $\quad T'(p) \circ T'(p^r) = T'(p^{r+1}) + p^{1-k}\ T'(p^{r-1}).$

3. An easy induction, and the combination of 1. and 2. then yields:

$$(*) \qquad T'(n)\,T'(m) = \sum_{d|n,m} d^{1-k}\, T'\left(\frac{nm}{d^2}\right).$$

Define the Hecke-operators $T(n) = n^{k-1}\,T'(n)$. The identity (*) becomes:

$$T(n)\,T(m) = \sum_{d|n,m} d^{k-1}\, T\left(\frac{nm}{d^2}\right).$$

Hence, the Hecke-operators generate a commutative ring. The identities relating the T(n) are fully expressed by the following product expansion for the formal Dirichlet series:

$$\sum_{n=-\infty}^{+\infty} T(n)n^{-s} = \Pi_p\,(1 - T(p)p^{-s} + p^{k-1-2s})^{-1}.$$

We now use the inhomogeneous notation. Let M(n) be the set of integral matrices of determinant n. We have $M(n) = \cup_i\, \Gamma'\,\alpha_i$ (disjoint), and there is a one-to-one correspondence between sublattices $L' \subset L$ with $[L : L'] = n$ and lattices $\alpha_i(L)$. From the correspondence $f|\alpha \to |\alpha|^{k/2}\, F \circ \alpha$ we get the inhomogeneous action of T(n):

$$f|T(n) = n^{\frac{k}{2}-1} \sum_i f|\alpha_i,$$

and the same identity (*) holds here.

PROPOSITION 5: If $f \in M(k)$, $S(k)$ resp., then $f|T(n) \in M(k)$, $S(k)$ resp..

PROOF:
$$M(n) = \bigcup_{\substack{ad = n\\ d > 0\\ b \bmod d}} \Gamma' \begin{pmatrix} a & b\\ 0 & d\end{pmatrix} \quad \text{(disjoint)}.$$

Let $f(\tau) = \sum_{\nu=0}^{\infty} a_\nu\, e^{2\pi i\nu\tau} \in M(k)$. By definition,

$$f|T(n)(\tau) = n^{k-1} \sum_{\substack{ad = n\\ d > 0\\ b \bmod d}} d^{-k}\, f\left(\frac{a\tau+b}{d}\right)$$

$$= n^{k-1} \sum_{d|n} d^{-k} \sum_{\nu=0}^{\infty} a\nu \sum_{b \bmod d} e^{2\pi i\nu\frac{a\tau}{d}}\, e^{2\pi i\frac{\nu b}{d}}$$

$$= \sum_{d|n} a^{k-1} \sum_{\nu=0}^{\infty} a_{\frac{\nu n}{d}}\, q^{\nu d} \qquad \left(\text{with } a = \frac{n}{d}\right)$$

$$= \sum_{\nu=0}^{\infty} q^{\nu} \left(\sum_{d|n,\nu} d^{k-1} a_{\frac{n\nu}{d^2}}\right),$$

$$f|T(n)(\tau) = \sum_{\nu=0}^{\infty} a_{\nu}(n)\, q^{\nu} \quad \text{with}$$

$$a_{\nu}(n) = \sum_{d|\nu,n} d^{k-1} a_{\frac{\nu n}{d^2}},$$

and this entails that $f|T(n)$ is a modular form (resp. cusp form) if f is. Now, if we take $f(\tau)$ to be an eigenfunction for all $T(n)$, i.e.,

$$f|T(n) = c(n)\, f,$$

then $a_1\, c(n) = a_1(n) = a_n$, hence the Fourier coefficients and the eigenvalues of an eigenfunction are proportional. We normalize the eigenfunction f such that $a_1 = 1$. Now, let $f \not\equiv 0$ be a normalized eigenfunction for all Hecke-operators, then $a_n = c(n)$ and from the identity for the formal Dirichlet-series we get an Euler-product for the eigenfunctions of the $T(n)$:

$$\phi_f(s) = \sum_{n=1}^{\infty} a_n\, n^{-s} = \Pi_p\, (1-a_p\, p^{-s} + p^{k-1-2s})^{-1}$$

Example: $$\Delta(\tau) = q\, \Pi_{n=1}^{\infty}\, (1-q^n)^{24} = \sum_{n=1}^{\infty} \tau(n)\, q^n$$

$$\phi_{\Delta}(s) = \sum_{n=1}^{\infty} \tau(n)\, n^{-s}.$$

(The product for $\phi_{\Delta}(s)$ was first proved by Mordell (1917)).
The space of cusp-forms of weight 12 has dimension 1, so $\Delta(\tau)$ is an eigenform for all $T(n)$ by the foregoing theorem, hence we get the Euler-product

$$\phi_{\Delta}(s) = \Pi_p\, (1-\tau(p)\, p^{-s} + p^{11-2s})^{-1}.$$

Ramanujan conjectured $|\tau(p)| \leq 2p^{11/2}$, which is equivalent with $0 = 1 - \tau(p)t - p^{11}\, t^2$, having conjugate roots. The fact that a modular form f is an eigenfunction for $T(p)$ is related to the existence of a p-Euler-product for the associated Dirichlet series, as follows:

Ogg-20

PROPOSITION 6: If $f(\tau)$ is an eigenfunction for $T(p)$, then $\phi(s) = \phi_f(s)$ has a p-Euler-product, i.e.,

$$\phi(s) = \sum_{n=1}^{\infty} a_n\, n^{-s} = \left(\sum_{(m,p)=1} a_m\, m^{-s}\right)\left(\sum_{\nu=0}^{\infty} c(p^{\nu})\, p^{\nu s}\right)$$

and vice-versa.

PROOF: We use the following lemma:

LEMMA: $f \in M(k)$ and $f(\tau+\frac{1}{p}) = f(\tau) \Rightarrow f = 0$.

Indeed $\begin{pmatrix} p & 1 \\ 0 & p \end{pmatrix}$ is a primitive matrix of determinant p^2 and, as such, $\begin{pmatrix} p & 1 \\ 0 & p \end{pmatrix} \in \Gamma' \begin{pmatrix} 1 & 0 \\ 0 & p^2 \end{pmatrix} \Gamma'$, $\gamma\begin{pmatrix} p & 1 \\ 0 & p \end{pmatrix}\gamma' = \begin{pmatrix} 1 & 0 \\ 0 & p^2 \end{pmatrix}$, say. Then, $f|\gamma \begin{pmatrix} p & 1 \\ 0 & p \end{pmatrix}\gamma' = f|\begin{pmatrix} 1 & 0 \\ 0 & p^2 \end{pmatrix}$, and $f|\gamma \begin{pmatrix} p & 1 \\ 0 & p \end{pmatrix}\gamma' = f|\begin{pmatrix} p & 1 \\ 0 & p \end{pmatrix}$, implying $f(\tau) = f(\tau/p^2)$. Calculating the Fourier coefficients of f we get $f = 0$.

Now suppose $\phi(s)$ has a p-Euler-product, i.e.,

$$a_{pm} = c(p)\, a_m, \qquad p \nmid m.$$

We get, $f|T(p) - c(p)\, f(\tau) = p^{k-1} \sum_{\nu=0}^{\infty} a_\nu q^{p\nu} + \sum_{\nu=0}^{\infty} a_{p\nu} q^{\nu} - c(p)\sum_{\nu=0}^{\infty} a_\nu q^{\nu}$.

Hence, $(f|T(p) - c(p)f)(\tau)$ is a power series in q^p and thus zero by the lemma. It is clear that $f|T(p) = c(p)f$, entails that the p-factor in the product is necessarily equal to $(1 - c(p)\, p^{-s} + p^{k-1-2s})^{-1}$.

COROLLARY: If $f|T(n) = c(n)f$ for all n, f being normalized, then:

$$\phi_f(s) = \Pi_p\, (1 - c(p)\, p^{-s} + p^{k-1-2s})^{-1}.$$

In the decomposition $M(k) = \mathbf{C}\, E_k(\tau) \oplus S(k)$ we see that the Dirichlet series $\zeta(s)\, \zeta(s+1-k)$, corresponding to $E_k(\tau)$, has an Euler-product, hence, $E_k(\tau)$ is an eigenfunction for all Hecke-operators. This means that $T(n)$ respects the decomposition. Furthermore, it is possible to diagonalize all $T(n)$ on $M(k)$ (and to do this it will be sufficient to do it on $S(k)$) using an inner-product, defined by H.Petersson, for which the $T(n)$ became Hermitian operators.

REMARKS:

1. If $f \in S(k)$ then $|y^{k/2}\ f(\tau)|$ is Γ-invariant and vanishes at ∞, hence it is bounded in the fundamental domain and therefore bounded in H. We have $f(\tau) = O(y^{-k/2})$ and

$$a_n = \int_0^1 e^{-2\pi i n(x+iy)}\ f(x+iy)dx = O(n^{k/2}),$$

so $\phi_f(s)$ converges for $R(s) > \frac{k}{2} + 1$.

2. The eigenvalues of the Hecke-operators are algebraic integers. First we prove inductively that the $\mathbb{Z}$-module of modular forms with integral Fourier coefficients, $M(k)_{\mathbb{Z}}$, is a free $\mathbb{Z}$-module of rank n = dim $M(k)_{\mathbb{C}}$. Given $f(\tau) = \sum_{n=0}^{\infty} a_n q^n$ with $a_n \in \mathbb{Z}$, write $f = a_0\ E_4^a\ E_6^b + \Delta g$; with $4a + 6b = k$, and $g \in M(k-12)$. Then $a_0 \in \mathbb{Z}$, g has Fourier coefficients in $\mathbb{Z}$, and by induction $M(k-12)_{\mathbb{Z}}$ is free of rank n-1 = dim $M(k-12)$. Hence we may find a basis for $M(k)_{\mathbb{C}}$ such that the T(n) are represented by integral matrices and so the characteristic polynomials have integral coefficients.

5. Forms for subgroups of the modular group

Let G be a subgroup of finite index μ in Γ. The fundamental domain of Γ is $D = D(\Gamma)$. We have a disjoint union $H = \cup_{\gamma \in \Gamma}\ \gamma\ D(\Gamma)$, and $\Gamma = \cup_{i=1}^{\mu}\ G\ \alpha_i$, hence $D(G) = \cup_{i=1}^{\mu}\ \alpha_i\ D(\Gamma)$. For the invariant measure $dxdy/y^2$ in H , we have that the proportion, measure of D(G) / measure of $D(\Gamma)$, is equal to $\mu = [\Gamma : G]$. Let $\alpha \in SL(2, \mathbb{R})/\pm I$ such that $\alpha\ G\ \alpha^{-1} \subset \Gamma$, then $\alpha D(G)$ is a fundamental domain for $\alpha G \alpha^{-1}$ with the same measure as D(G), therefore $[\Gamma : \alpha\ G\ \alpha^{-1}] = [\Gamma : G]$. Concerning the analytic structure of H/G and the holomorphicity of $H \to H/G \to H/\Gamma$ there are only problems in the cusps. At ∞; let e be the least positive integer such that $\begin{pmatrix} 1 & e \\ 0 & 1 \end{pmatrix} \in G$, we take $q = e^{2\pi i \tau/e}$ as a local variable at ∞. If $P = \gamma^{-1}(\infty)$ is another cusp, $\gamma \in \Gamma$, we throw P to infinity by γ and $f(\tau)$ is said to be holomorphic at $P = \gamma^{-1}(\infty)$ if and only if $f \circ \gamma$ (on $H/\gamma^{-1}\ G\gamma$) is holomorphic at ∞ as above.

Ogg-22

DEFINITION: <u>A modular form of weight k for G</u> is a holomorphic function $f(\tau)$ on H such that

1. $f|_k\ \alpha = f$ for all $\alpha \in G$
2. $f|_k\ \gamma$ is holomorphic at ∞ for all $\gamma \in \Gamma$.

A form is said to be a <u>cusp form for G</u> if $f|_k\ \gamma$ vanishes at ∞ for all $\gamma \in \Gamma$. Moreover, $f(\tau)$ has weight 2 for G if and only if $f(\tau)d\tau$ is G-invariant. At ∞ we have $f(\tau)d\tau = \sum_\nu a_\nu q^\nu \frac{2\pi i}{e}\frac{dq}{q}$, and if $f(\tau)$ is to be a cusp form, the pole at $q = 0$ has to cancel out; hence the zero-term should be missing. That means then that $f(\tau)d\tau$ is a holomorphic differential.

PROPOSITION 7: <u>If</u> $f \in M(k,G)$ <u>then</u> f <u>has</u> $\frac{k\mu}{12}$ <u>zeroes</u> (<u>measured in local variables</u>) <u>in the fundamental-domain</u>.

PROOF: f^{12}/Δ^k is a function on H/G and, as such, it has as many zeroes as poles, but f^{12}/Δ^k has exactly $k\mu$ poles in the fundamental domain, hence f has $\frac{k\mu}{12}$ zeroes in the fundamental domain.

COROLLARY: $\dim M(G,k) \leq 1 + [\frac{\mu k}{12}]$, a better bound can be found, using the Riemann-Roch theorem (cf. Shimura's book).

6. The Petersson product

Put $\Gamma(N) = \{\begin{pmatrix} a & b \\ c & d\end{pmatrix} \equiv \begin{pmatrix} 1 & 0 \\ 0 & 1\end{pmatrix} \bmod N\}$. A modular form (cusp form) of <u>level N</u> is a modular form (cusp form) for $G = \Gamma(N)$.

LEMMA 4: <u>Let</u> $\beta \in M(n)$, <u>then</u>

$$\beta\ \Gamma(N)\ \beta^{-1} \supset \Gamma(nN).$$

From this it follows that:

PROPOSITION 8: <u>If</u> f <u>is a form of level</u> N <u>then</u> $f|\beta$ <u>has level</u> nN <u>and</u>

$f|\beta$ _is a cusp form if_ f _is_.

Let $f(\tau)$, $g(\tau)$ be cusp forms of some unspecified, but fixed, level N. Let $\tau = x+iy$, $\begin{pmatrix} a & b \\ c & d \end{pmatrix} \in SL(2, \mathbb{R})$ and $\tau' = \frac{a\tau+b}{c\tau+d} = x'+iy'$. Then, $y' = y/|c\tau+d|^2$. We put $\delta(f,g) = f(\tau)\overline{g(\tau)}\, y^k \frac{dx\, dy}{y^2}$. It is clear that for any $\alpha \in SL(2, \mathbb{R})$ we have:

$$\delta(f,g) \circ \alpha = f\left(\frac{a\tau+b}{c\tau+d}\right) \overline{g\left(\frac{a\tau+b}{c\tau+d}\right)} \frac{y^k}{|c\tau+d|^{2k}} \frac{dx\, dy}{y^2}$$

$$= \delta(f|\alpha, g|\alpha).$$

The inner product is defined as follows:

$$(f,g) = \int_{D(N)} \delta(f,g) \times 1/\,[\Gamma:\Gamma(N)].$$

$$= \frac{1}{\mu} \int_{D(N)} f(\tau)\, \overline{g(\tau)}\, y^{k-2}\, dx\, dy$$

(where $D(N) = D(\Gamma(N))$ is the fundamental domain for $\Gamma(N)$. Note that all this is independent of the choice of the level N of f and g. It remains to be checked whether the integral converges or not. At ∞ we have that $f(\tau)\overline{g(\tau)} = O(e^{-cy})$, $c > 0$, f and g being cusp forms. At the other cusps we proceed in the same way after throwing the cusp to ∞.

PROPOSITION 9: _If_ $\beta \in M(n)$ _and if_ f, g _are cusp-forms_, _then_ $(f|\beta, g) = (f, g|\beta^{-1})$.

PROOF: Since $\beta \in M(n)$,

$$(f|\beta, g|\beta) = \frac{1}{[\Gamma:\Gamma(N)]} \int_{D(N)} \delta(f|\beta, g|\beta)$$

$$= \frac{1}{[\Gamma:\Gamma(N)]} \int_{D(N)} \delta(f,g) \circ \beta$$

$$= \frac{1}{[\Gamma:\Gamma(N)]} \int_{\beta D(N)} \delta(f,g),$$

where $\beta D(N)$ is a fundamental domain for the conjugate group $\beta\Gamma(N)\beta^{-1}$ with index $[\Gamma:\Gamma(N)]$, so,

$$(f|\beta,\ g|\beta) = (f,g),$$

proving our assertion. Writing, $M(p) = \Gamma' \begin{pmatrix} 1 & 0 \\ 0 & p \end{pmatrix} \Gamma' = \cup_i \Gamma' \alpha_i$, the Hecke-operators $T(p)$ are given by

$$T(p) = p^{k/2\ -1} \sum_i \alpha_i.$$

Since, in a single double coset, each right coset meets each left coset, we may choose representatives $\{\alpha_i\}$ such that :

$$M(p) = \cup_i \Gamma' \alpha_i = \cup_i \alpha_i \Gamma'.$$

Denote $\begin{pmatrix} a & b \\ c & d \end{pmatrix}' = \begin{pmatrix} d & -b \\ -c & a \end{pmatrix}$. Then $M(p) = \dot{\cup}_i \Gamma' \alpha_i = \cup \Gamma' \alpha_i'$ entails,

$$T(p) = p^{k/2\ -1} \sum_i \alpha_i = p^{k/2\ -1} \sum_i \alpha_i'.$$

For $f,\ g \in S(k, \Gamma)$ this yields

$$\begin{aligned}(f|T(p),\ g) &= p^{k/2\ -1} \sum_i (f|\alpha_i,\ g) \\ &= p^{k/2\ -1} \sum (f,\ g|\alpha_i') \\ &= (f,\ g|T(p))\end{aligned}$$

which proves:

THEOREM 5: <u>The Hecke-operators are Hermitian operators on</u> $S(k)$; <u>they generate a commutative ring of Hermitian operators and so</u>, <u>the space of cusp forms allows a basis of simultaneous eigenfunctions for all</u> $T(n)$; <u>they may be normalized to have leading coefficient</u> 1. <u>Since</u> $M(k) = \mathbb{C}\, E_k \oplus S(k)$, <u>the Hecke-operators are diagonalized in</u> $M(k)$.

7. Eisenstein series of higher level

For the homogeneous principal congruence subgroup $\Gamma'(N)$ we have an exact sequence:

$$1 \to \Gamma'(N) \to SL(2, \mathbb{Z}) \to SL(2, \mathbb{Z}/n\,\mathbb{Z}) \to 1$$

which entails: $\mu(N) = [\Gamma:\Gamma(N)] = \frac{N^3}{2} \Pi_{p|N} (1-p^{-2})$ for $N \geq 3$. The Riemann surface will be $X(N) = H/\Gamma(N) \cup \{\text{cusps}\}$. At ∞; $\binom{1\ N}{0\ 1}$ is the least translation in $\Gamma'(N)$ and hence the appropriate local variable is $q^{1/N}$. The ramification-index e_∞ of ∞ in $X(N) \to X = \widehat{H/\Gamma}$, is exactly N, and since $\Gamma(N)$ is a normal subgroup of Γ, all cusps have the same ramification index N. Thus, the number $\sigma(N)$ of cusps for $\Gamma(N)$ is equal to $\frac{\mu(N)}{N} = \frac{N^2}{2} \Pi_{p|N} (1-p^{-2})$. It is useful to denote a cusp by a column $\pm\binom{x}{y}$, with x, y mod N, and $(x, y, N) = 1$; in fact, the cusp then corresponds with $x/y \in \mathbb{Q} \cup \{\infty\}$, e.g. $\infty = \binom{1}{0}$. This notation has been chosen because it gives exactly the cusps, unequivalent under the action of $\Gamma(N)$. Further, $SL(2, \mathbb{Z})$ operates on the left $\begin{pmatrix} a & b \\ c & d \end{pmatrix} \binom{x}{y}$, and if $\Gamma \supset G \supset \Gamma(N)$, then a cusp for G is an orbit of $\pm\binom{x}{y}$ under the action of $G/\Gamma(N)$.

DEFINITION: Let $E = \mathbb{C}/L$ be an elliptic curve. To a point, $a_1 \frac{\omega_1}{N} + a_2 \frac{\omega_2}{N}$, a_1, a_2 mod N $(a_1,a_2) \not\equiv (0,0)$, there corresponds a <u>Teilwert</u>, i.e.,

$$\wp\left(\frac{a_1\omega_1 + a_2\omega_2}{N}, \omega_1, \omega_2\right),$$

a function of ω_1, ω_2 which is parametrized by a_1, a_2. It is sometimes written $f(a_1, a_2, \omega_1, \omega_2)$. From the definition it follows that:

$$f(a_1, a_2, \omega_1, \omega_2)$$

$$= \frac{1}{\left(\frac{a_1\omega_1+a_2\omega_2}{N}\right)^2} + \sum_{n_1,n_2 \in \mathbb{Z}} \frac{1}{\left(\frac{a_1\omega_1+a_2\omega_2}{N} + n_1\omega_1+n_2\omega_2\right)^2} - \frac{1}{(n_1\omega_1+n_2\omega_2)^2}$$

$$= \sum_{n_1\equiv a_1(N)} \sum'_{n_2\equiv a_2(N)} \frac{N^2}{(n_1\omega_1+n_2\omega_2)^2} - G_2(\omega_1, \omega_2)$$

$$= G_2(a_1, a_2, N, \omega_1, \omega_2) - G_2(\omega_1, \omega_2),$$

the last equality defining the function $G_2(a_1, a_2, N, \omega_1, \omega_2)$.

Obviously: $f(a_1, a_2, \omega_1, \omega_2) \circ \begin{pmatrix} a & b \\ c & d \end{pmatrix} = f((a_1, a_2)\begin{pmatrix} a & b \\ c & d \end{pmatrix}, \omega_1, \omega_2)$ and in particular, if $\begin{pmatrix} a & b \\ c & d \end{pmatrix} \in \Gamma(N)$ then $f(a_1, a_2, \omega_1, \omega_2)$ is invariant

Ogg-26

for $\begin{pmatrix} a & b \\ c & d \end{pmatrix}$. The difference $G_2(a_1, a_2, \omega_1, \omega_2) - G_2(\omega_1, \omega_2)$ is a modular form of weight 2 and level N, (convergence assured under good conditions). Eisenstein series, $G_k(\alpha, \beta, N, \tau)$, of level N and with $k > 2$ are defined by:

$$G_k(\alpha, \beta, N, \tau) = \sum_{n\equiv\alpha(N)} \sum'_{m\equiv\beta(N)} \frac{N^k}{(n\tau+m)^k} \times \left(\frac{1}{2\pi i}\right)^k .$$

This series converges absolutely for $k > 3$ and is then a modular form of weight k and level N, with the property:

$$G_k(\alpha, \beta, N, \tau) \mid \begin{pmatrix} a & b \\ c & d \end{pmatrix} = G_k((\alpha, \beta) \begin{pmatrix} a & b \\ c & d \end{pmatrix}, N, \tau).$$

If $\begin{pmatrix} a & b \\ c & d \end{pmatrix} \in \Gamma(N)$ we have,

$$G_k(\alpha, \beta, N, \tau) \mid \begin{pmatrix} a & b \\ c & d \end{pmatrix} = G_k(\alpha, \beta, N, \tau),$$

and to see that it is a modular form it is sufficient to compute the Fourier-series:

$$G_k(\alpha, \beta, N, \tau) = \delta\left(\frac{\alpha}{N}\right) \left(\frac{N}{2\pi i}\right)^k \sum_{m\equiv\beta\ (N)}' m^{-k}$$

$$+ \sum_{n\equiv\alpha\ (N)}' \sum_{mn > 0} \frac{(-1)^k}{(k-1)!} m^{k-1} \operatorname{sgn}(m) e^{\frac{2\pi i m(n\tau+\beta)}{N}}$$

with $\delta\left(\frac{\alpha}{N}\right) = \begin{cases} 0 \text{ if } \alpha \not\equiv 0 \bmod N \\ 1 \text{ if } \alpha \equiv 0 \bmod N. \end{cases}$

We see that this is holomorphic at ∞ and hence also at all other cusps. It is possible to construct similar functions in case $k = 1$, (Hecke). Assume, $(\alpha, \beta, N) = 1$. The number of cusps $\frac{N^2}{2} \prod_{p|N} (1 - p^{-2})$ is then also the number of series constructed, since $G_k(-\alpha, -\beta, N, \tau) = (-1)^k G_k(\alpha, \beta, N, \tau)$. Introducing so-called restricted Eisenstein-series, one can find elements of $E(N, k)$, the space generated by all $G_k(\alpha, \beta, N, \tau)$, with value 1 at a chosen cusp and 0 at the other cusps. Those forms are therefore independent and, for $k > 3$,

$$M(N, k) = E(N, k) \oplus S(N, k).$$

The theory of Hecke-operators $T(n)$ with $(n, N) = 1$, and the Petersson inner product can be expounded in level N too.

As earlier, $T(n)$, $(n, N) = 1$, preserves the decomposition of $M(N, k)$.

Fourier expansion of $G_k(\alpha, \beta, N, \tau)$ yields that this function has a zero at ∞ of order $\{\alpha\}$, $0 < \{\alpha\} < N/2$ where $\{\alpha\} \equiv \pm\alpha \pmod N$.

$\Gamma(N)$ being normal in Γ, we deduce that $G_k(\alpha, \beta, N, \tau)$ has a zero of order $\{\alpha x + \beta y\}$ at the cusp $\pm\binom{x}{y}$. Easy calculation shows that, whenever $(\alpha, \beta, N) = 1$, $G_k(\alpha, \beta, N, \tau)$ has a total of $\frac{\mu(N)}{4} = \frac{N^3}{8} \Pi_{p|N} (1 - p^{-2})$ zeroes at the cusps. In case $k = 3$, no other zeroes are possible and therefore $G_3(\alpha, \beta, N, \tau)$ has all its zeroes situated at the cusps, and there are exactly $3\mu(N)/12$ zeroes.

We introduce now, a larger congruence subgroup $\Gamma_0(N)$ of Γ,

$$\Gamma_0(N) = \{\begin{pmatrix} a & b \\ c & d \end{pmatrix} \equiv \begin{pmatrix} * & * \\ 0 & * \end{pmatrix} \bmod N\}.$$

The cusps are given by $\begin{pmatrix} a & b \\ 0 & c \end{pmatrix} \binom{x}{y}$. $H_N = \begin{pmatrix} 0 & -1 \\ N & 0 \end{pmatrix}$ normalizes $\Gamma_0(N)$ because

$$H_N \begin{pmatrix} a & b \\ c & d \end{pmatrix} H_N^{-1} = \begin{pmatrix} d & -c/N \\ -bN & a \end{pmatrix}$$

and hence, H_N simply permutes the cusps.

Let N be a prime p, then we have cusps $\binom{1}{0} = \infty$ and $\binom{0}{1} = 0$. At $\binom{1}{0}$ the ramification index under $X_0(p) \to X$ equals 1, at $\binom{0}{1}$ it is p. The operator $H_p = \begin{pmatrix} 0 & -1 \\ p & 0 \end{pmatrix}$ exchanges the cusps and $[\Gamma:\Gamma_0(p)] = p+1$. For $\Gamma_0(p)$, $p > 3$, one can construct modular forms using the Eisenstein series.

Let f be the function $\Pi_{j=1}^{\frac{p-1}{2}} G_3(0, j)$. $\Gamma_0(p)$ acts on pairs $(0, j)$ by:

$$(0, j) \begin{pmatrix} a & b \\ 0 & d \end{pmatrix} = (0, dj).$$

PROPOSITION 10: The function f is a modular form of weight $k = \frac{3}{2}(p-1)$ for $\Gamma_0(p)$, with character $\varepsilon(d) = (\frac{d}{p})$, the Legendre symbol; f is not vanishing at ∞ and f^2 has

$$\frac{\mu k}{12} = \frac{3\mu(p-1)}{24} = \frac{p^2 - 1}{4} \quad \text{zeroes at } 0.$$

Moreover, $\Delta_p(\tau) = \Delta(p\tau)$ is clearly a cusp form of weight 12 for $\Gamma_0(p)$. If we then consider Δ^p/Δ_p, the quotient of two cusp forms, we get a modular form for $\Gamma_0(p)$ of weight $(p-1)/12$, with no poles and with a zero of order $p^2 - 1$ at the cusp 0. On the other hand,

$$\frac{\Delta^p}{\Delta_p} \mid \begin{pmatrix} 0 & 1 \\ p & 0 \end{pmatrix} = \frac{\Delta_p^p}{\Delta},$$

and Δ_p^p/Δ has a zero of order $p^2 - 1$ at ∞. It follows that $f^8 = (\mathrm{cst})\ \Delta^p/\Delta_p$ or,

$$f = (\mathrm{cst})\ (\eta^p/\eta_p)^3.$$

In a similar way, using G_2's, one finds that $(\frac{\eta^p}{\eta_p})^2$ is a form for $\Gamma_0(p)$; it is even a form without character, and η^p/η_p is then a modular form of weight $k = \frac{p-1}{2}$ for $\Gamma_0(p)$ with character $(\frac{d}{p})$, which proves:

THEOREM 6 (Hecke): η^p/η_p <u>is a modular form of weight</u> $\frac{p-1}{2}$ <u>for</u> $\Gamma_0(p)$, <u>with character</u> $(\frac{d}{p})$.

We are now in a position to say more about the cusps of $\Gamma_0(p)$.

EXAMPLE: Put, $\eta(\tau) = q^{1/24}\Pi_{n=1}^{\infty}(1-q^n)$. Take $p = 11$; $X_0(11)$ has genus 1. We see that η_{11}/η^{11} is a modular form such that the corresponding character can be eliminated by squaring the form, and

$$\Delta\ (\frac{\eta_{11}}{\eta^{11}})^2 = \eta(\tau)^2\ \eta(11\tau)^2 = f,$$

is a cusp form of weight 2. We have

$$f(\tau) = q\ \Pi_{n=1}^{\infty}\ (1 - q^n)^2\ (1 - q^{11n})^2$$
$$= \textstyle\sum_{n=1}^{\infty} a_n\ q^n,$$

and a corresponding series $\phi_f(s) = \sum_{n=1}^{\infty} a_n\ n^{-s}$ which, by Hecke-theory, has an Euler-product:

$$\phi_f(s) = (11)^{-s}\ \Pi_{p\nmid 11}\ (1 - a_p\ p^{-s} + p^{11-2s})^{-1}.$$

The latter Dirichlet series will turn out to be the zeta-function of the elliptic curve $X_0(11)$ (cf. Shimura's book). Moreover, in this case we have $|a_p| \leq 2\, p^{1/2}$, by the "Riemann hypothesis" for elliptic curves. Now, we use η^p/η_p to cook up functions for $\Gamma_0(p)$.
Let $h \mid (p-1,\ 12)$ and put,

$$f(\tau) = \left(\frac{\eta(\tau)}{\eta(p\tau)}\right)^{24/h}$$

$$= \left(\frac{\eta^p}{\eta_p}\right)^{24/h} \Delta^{-\left(\frac{p-1}{h}\right)} .$$

Since $24/h$ is divisible by two, we eliminated the character and so $f(\tau)$ is a modular function on $X_0(p)$ with divisor $\frac{p-1}{h}\,((0) - (\infty))$, concentrated on the cusps. Conversely, suppose

$$f(\tau) = \left(\frac{\Delta(\tau)}{\Delta(p\tau)}\right)^{1/h}$$

is a function on $X_0(p)$, for some $h > 0$. Then

$$f(\tau) = q^{-\left(\frac{p-1}{h}\right)} \prod_{n=1}^{\infty} \left(\frac{1-q^n}{1-q^{pn}}\right)^{24/h} ,$$

and, as $f(\tau)$ is $\Gamma_0(p)$-invariant, we have $h|p-1$ and $f(\tau) \in \mathbb{Q}((q))$.
Write

$$\Gamma = \cup_{i=0}^{p} \Gamma_0(p)\, \alpha_i, \quad \alpha_0 = I \quad \text{and}$$

$$\alpha_i = \begin{pmatrix} 0 & -1 \\ 1 & i \end{pmatrix} \quad \text{for} \quad 1 \leq i \leq p.$$

The function $g(\tau) = \sum_i f(\alpha_i(\tau))$ is invariant under Γ and holomorphic in H (because η does not vanish in H), therefore $g(\tau)$ is a polynomial in $j(\tau)$. If $\nu \geq 1$,

$$f\left(\frac{-1}{\tau+\nu}\right) = \left[\frac{\Delta\left(\frac{-1}{\tau+\nu}\right)}{\Delta\left(\frac{-1}{(\tau+\nu)p}\right)}\right]^{1/h}$$

$$= \left[\frac{(\tau+\nu)^{12}\ \Delta(\tau+\nu)}{\left(\frac{\tau+\nu}{p}\right)^{12}\ \Delta\left(\frac{\tau+\nu}{p}\right)}\right]^{1/h} \qquad \text{and}$$

$$f(\frac{-1}{\tau+\nu}) = p^{12/h} f(\frac{\tau+\nu}{p})^{-1}.$$

Thus, $g(\tau) = f(\tau) + p^{12/h} \sum_{\nu=1}^{p} f(\frac{\tau+\nu}{p})^{-1} = q^{-(\frac{p-1}{h})} + \text{rest},$

where "rest" is in $\mathbb{Q}((q))$ and vanishes at ∞, since the pole at ∞ is transformed into a zero for f^{-1}. The fact that $g \in \mathbb{Q}((q))$ entails $p^{12/h} \in \mathbb{Q}$, i.e. $h|12$ and also $h|p-1$. Finally,

$$((\frac{\eta}{\eta_p})^{24/h}) = \frac{p-1}{h}((0) - (\infty))$$

yields the desired divisor equality. A similar result can be obtained if $N = p^2$, one finds then,

$$(\frac{\eta}{\eta_{p^2}}) = \frac{p^2-1}{24}((0) - (\infty)).$$

Note that $\frac{p-1}{h}$ resp. $\frac{p^2-1}{24}$ is the exact order of the divisor class.

COROLLARY: $\Gamma_0(p)$ <u>has genus</u> 0, <u>if and only if</u> $p = 2, 3, 5, 7, 13$.

Let $M(N, k, \varepsilon)$ be the space of functions f with:

$$f \mid \begin{pmatrix} a & b \\ c & d \end{pmatrix} = \varepsilon(d)f \quad \text{for} \quad \begin{pmatrix} a & b \\ c & d \end{pmatrix} \in \Gamma_0(N).$$

We define Hecke-operators as before, except we insist that the representatives of cosets in the double coset be $\equiv \begin{pmatrix} 1 & * \\ 0 & n \end{pmatrix} \bmod N$. Thus

$$T(n) = n^{k/2 - 1} \sum \gamma_i \alpha_i,$$

such that, if $\alpha_i \equiv \begin{pmatrix} a & * \\ 0 & * \end{pmatrix} \bmod N$, then

$$\gamma_i = \begin{pmatrix} a^{-1} & * \\ 0 & a \end{pmatrix} \bmod N,$$

i.e., working on functions, the multiplication by γ_i stands for the multiplication by $\varepsilon(a)$. This is possible if $(a, N) = 1$, but otherwise there is nothing to do because then the term would be skipped.

Hence, $$T(n) = n^{k/2 - 1} \sum_{\substack{ad = n \\ b \bmod d \\ d > 0}} \varepsilon(a) \begin{pmatrix} a & b \\ 0 & d \end{pmatrix} = T(n, N, \varepsilon).$$

Particularly, if n = p:

$$f \mid T(p)\ (\tau) = \varepsilon(p)\ p^{k-1}\ f(p\tau) + \frac{1}{p} \sum_{1 \bmod p} f\left(\frac{\tau+1}{p}\right)$$

$$= \varepsilon(p)\ p^{k-1} \sum_\nu a_\nu q^{p\nu} + \sum_\nu a_{p\nu} q^\nu .$$

If $p|N$ then $\varepsilon(p) = 0$, so T(p) exchanges the coefficient a_ν for $a_{p\nu}$.

We have:

1. $$T(n)\ T(m) = \sum_{d|n,m} \varepsilon(d)\ d^{k-1}\ T\left(\frac{nm}{d^2}\right)$$

and $$\sum_{n=1}^{\infty} T(n)\ n^{-s} = \Pi_p\ (I - T(p)\ p^{-s} + \varepsilon(p)\ p^{k-1-2s})^{-1}$$

2. $$f \mid T(n)\ (\tau) = \sum_{\nu=0}^{\infty} q^\nu \left(\sum_{d|n,\nu} \varepsilon(d)\ d^{k-1}\ a_{\frac{n\nu}{d^2}}\right)$$

Passing back and forth between $f(\tau)$ and $\phi_f(s)$ gives results which are similar to the results obtained in the level 1 case.

THEOREM 7: f _is a (normalized) eigenfunction for all Hecke-operators if and only if_

$$\phi_f(s) = \Pi_p\ (1 - a_p\ p^{-s} + \varepsilon(p)\ p^{k-1-2s})^{-1}.$$

We consider only cusp forms from now on, and try to expound the expected properties for the inner product.

It shows that Hecke-operators are ε-Hermitian, i.e.,

$$(f \mid T(n), g) = \varepsilon(n)\ (f, g \mid T(n))$$

for (n, N) = 1. We have a basis $\{f_1, \ldots, f_r\}$ of $S(N, k, \varepsilon)$ which are simultaneous eigenfunctions for all T(n) with (n, N) = 1. This, however, cannot be done in a unique way without introducing the operators T(p) with $p|N$.

Ogg-32

We distinguish two cases:

1. The "best" case.

Here, ε is a primitive character modulo N, hence N is the exact level. We assume that we have diagonalized the Hecke-operators T(n), (n,N) = 1.

LEMMA 6 (Hecke):

If $$f(\tau) = \sum_{\nu=1}^{\infty} a_\nu q^\nu \in S(N, k, \varepsilon)$$

is such that $a_\nu = 0$ *whenever* $(\nu, N) = 1$ *then* $f = 0$.

PROOF: Put N = p (the general case is similar). The assumption is then that $f(\tau + 1/p) = f(\tau)$. For x, y arbitrary in $\mathbb{Z}$,

$$f \mid \begin{pmatrix} 1 & x/p \\ 0 & 1 \end{pmatrix} \begin{pmatrix} 1 & 0 \\ p & 1 \end{pmatrix} \begin{pmatrix} 1 & y/p \\ 0 & 1 \end{pmatrix} = f,$$

this yields, $$f = f \mid \begin{pmatrix} 1+x & \frac{x+y+xy}{p} \\ p & 1+y \end{pmatrix} = f|\gamma.$$

We choose x, y such that $\gamma \in \Gamma_0(N)$ and $\varepsilon(1+y) \neq 0, 1$. Then $f = \varepsilon(1+y)\, f$, and $f = 0$ follows.

Taking $f = f_1$, $f|T(n) = c(n)\, f$ if $(n, N) = 1$ and hence $c(n)\, a_1 = a_n$ for $(n, N) = 1$. Assume $a_1 = 1$, which is possible since $a_1 \neq 0$. One sees that T(n) and T(p) commute for (n, N) = 1, p|N. Therefore,

$$f \mid T(p)\, T(n) = c(n)\, f \mid T(p)$$

and by the unique assertions of the lemma it follows that $f \mid T(p) = c(p)\, f$. This way, we obtained the diagonalization of all T(n) on cusp-forms.

For p|N we can prove $|c(p)| = p^{\frac{k-1}{2}}$.

2. The opposite case (Atkin - Lehner).

Here ε = I mod N. It is readily verified that the uniqueness is false; indeed, two different forms for $\Gamma_0(N)$ may very well have the same eigenvalues for T(n), (n, N) = 1. This problem can be avoided if we introduce the concepts, "old" and "new" forms.

Observe first that, if $f \in M(1, k, I)$ is an eigenform for all $T(n)$, then

$$f \mid \begin{pmatrix} N & 0 \\ 0 & 1 \end{pmatrix} \in S(N, k) \quad \text{and}$$

$$\begin{pmatrix} N & 0 \\ 0 & 1 \end{pmatrix} T(n) = T(n) \begin{pmatrix} N & 0 \\ 0 & 1 \end{pmatrix} \quad \text{if } (N, n) = 1.$$

Since for these forms the Hecke-operators are the same for $\Gamma_0(N)$ or Γ, the eigenvalues of $f(\tau)$ and $f(N\tau)$ for $T(n)$ are both equal to $c(n)$ in case $(n, N) = 1$. Now, take m,d such that $md|N$ and let f be a form for $\Gamma_0(m)$. We call then $f(d\tau)$ an <u>old form for $\Gamma_0(N)$</u>, and $S^{old}(N, k)$ is the space generated by all the old forms $f(d\tau)$, $d|N$.

All $T(n)$ with $(n, N) = 1$ operate on the space of old forms because

$$T(n) \begin{pmatrix} d & 0 \\ 0 & 1 \end{pmatrix} = \begin{pmatrix} d & 0 \\ 0 & 1 \end{pmatrix} T(n),$$

hence, old forms are transformed into old forms.

We define $S^{new} = (S^{old})^{\perp}$ in $S(N, k)$. Then $T(n)$, $(n, N) = 1$, operates also in the space of new forms and we easily get an orthogonal basis for the space of new forms, consisting of simultaneous eigenforms for all $T(n)$ with $(n, N) = 1$. Moreover, it turns out that they are in fact eigenforms for all $T(n)$. The corresponding eigenvalues for $T(p)$, with $p|N$, are:

$$c(p) = \begin{cases} 0 & \text{if } p^2 \mid N \\ \pm p^{k/2-1} & \text{if } p \mid\mid N \text{, i.e., } p \text{ divides } N \text{ exactly.} \end{cases}$$

We mention some applications of the theory of $\Gamma_0(N)$-forms.

1). Theta-series

Let A be an even-integral matrix, i.e., $a_{ij} \in \mathbb{Z}$, $a_{ii} \in 2\mathbb{Z}$.

Let $Q(x) = \frac{1}{2} x^t A x$, be an integral, positive, quadratic form in an even number of variables. To $Q(x)$ we associate a Θ-series:

$$\Theta(\tau) = \sum_n e^{2\pi i \tau Q(n)},$$

and this provides us with modular forms of weight k and level N, where

N is the least positive integer for which $N A^{-1}$ is even-integral. (There are also more general theta-series, constructed by using spherical functions). The function $\Theta(\tau)$ is a form for $\Gamma_0(N)$ with character $\varepsilon(d) = (\frac{\Delta}{d})$, Δ being the discriminant of the quadratic form. In this way we derive certain cusp-forms from quadratic forms and the question arises: how many cusp-forms can be found this way? It is known that, in several cases, all cusp-forms can be derived from quadratic forms (Eichler).

When quadratic forms in an odd number of variables are being considered, then modular forms of half integral weight arise (see Shimura's lecture).

2). Weil's theorem on $\Gamma_0(N)$

Let $f(\tau)$ be a modular form for $\Gamma_0(N)$;

$$f(\tau) = \sum_{n=0}^{\infty} a_n q^n, \qquad \phi_f(s) = \sum_{n=1}^{\infty} a_n n^{-s}.$$

For $\Gamma_0(N)$ we need more than one functional equation to describe the relations between f and ϕ_f, because $\Gamma_0(N)$ has too many generators. Therefore, we introduce characters χ with conductor $t(\chi)$ such that $(t(\chi), N) = 1$, and we form

$$\phi_\chi (s) = \sum_{n=1}^{\infty} a_n (\chi(n)) n^{-s}.$$

Put
$$\Phi_\chi (s) = (\frac{2\pi}{t(\chi)})^{-s} \Gamma (s) \phi_\chi(s).$$

As a generalization of Hecke's theorem, Weil proves that f is a modular form for $\Gamma_0(N)$, if and only if $\Phi_\chi (k-s) = (\text{cst}) \Phi_\chi (s)$, i.e., we get a system of functional equations instead of only one; here the constant is a precise number involving Gauss sums, etc...

3). Let E be an elliptic curve defined over $\mathbb{Q}$. Let L(s) be the 1-dimensional part of the ζ-function of $E|\mathbb{Q}$ and let N be the conductor of E.

We have the Hasse[+] conjecture, saying that all series $L_\chi (s)$ satisfy functional equations as in 2. Weil conjectured that every ζ-function belonging to an elliptic curve E can be found as being a series associated to an element of $S(N, 2, 1)$.

REFERENCES

[1] A.O.L. ATKIN, J. LEHNER: Hecke Operators on $\Gamma_0(m)$, Math. Ann. 185, pp. 134-160 (1970).

[2] G.H. HARDY, M. RIESZ: The General Theory of Dirichlet's Series, Cambridge tracts in Math. and Math. Phys., vol.18, (1964).

[3] E. HECKE: Mathematische - Werke, Göttingen, Vandenhoeck & Ruprecht 1970.

[4] J.I. IGUSA: Theta Functions, Springer Verlag 1972.

[5] M. KNOPP: Modular Functions in Analytic Number Theory.

[6] J. LEHNER: Discontinuous Groups and Automorphic Functions, AMS, 1964.

[7] A. OGG: Modular Forms and Dirichlet Series, Benjamin 1969.

[8] J.P. SERRE: Cours d' Arithmetique, P.U.F.

[9] G. SHIMURA: Arithmetic Theory of Automorphic Functions, Publ. of the Math. soc., Japan (1971), no.11.

[10] A. WEIL: Über die Bestimmung Dirichletscher Reihen durch Funktionalgleichungen, Math. Ann. 168, pp. 149-156 (1967).

COMPLEX MULTIPLICATION

by Goro Shimura

(Notes by Willem Kuyk)

International Summer School on Modular Functions

Antwerp 1972

CONTENTS

The purpose of this article is to explain the main ideas of the topic with neither proofs, nor historical remarks. The exposition is divided into three parts. No modular functions appear in Part I except that the values of the j-function are needed as invariants of elliptic curves. Part II relies on the theory of Hecke operators (on cusp forms of weight 2), while Part III concerns the modular functions of arbitrary level. Almost nothing is new in the sense that most of the results, with proofs, can be found in [3], [4], [5], but I have attempted to present the material as accessibly (and perhaps dogmatically) as possible. The formulation in Part II is essentially equivalent to, but somewhat different from, what was done in [3, §7.5].

I would like to express my hearty thanks to W.Kuyk for preparing careful notes of my lectures.

Goro SHIMURA

PART I. COMPLEX MULTIPLICATION IN TERMS OF ELLIPTIC CURVES

§1. Short review of Class Field Theory

Let K be an algebraic number field, K_{ab} its maximal abelian extension (i.e. the composite of all finite extensions of K with abelian Galois group). Denote by K_A^* the group of idèles of K, and by $K_{\infty+}^*$ the identity component of the "archimedean part" of it. Viewing $K^* = K \setminus \{o\}$ as the subgroup of K_A^* of principal idèles, we have the fundamental exact sequence of class field theory :

$$(1) \quad 1 \to \overline{K^* K_{\infty+}^*} \to K_A^* \to \mathrm{Gal}(K_{ab}/K) \to 1,$$

where the bar denotes the closure with respect to the usual idèle topology on K_A^*. The p-th component of an element $x \in K_A^*$ is denoted by x_p, as usual. We signify by il(x) the ideal associated with the idèle x, i.e. the ideal satisfying the equality $\mathrm{il}(x)_p = x_p v_p$, for all p, where v_p denotes the maximal compact subring of the completion K_p of K at p. The image of $x \in K_A^*$ in $\mathrm{Gal}(K_{ab}/K)$ is denoted by $[x,K]$, and the reader is reminded of the fact that, if $x = (\ldots x_p, \ldots) \in K_A^*$ satisfies $x_p \equiv 1 \pmod{f}$, where f is the conductor of a finite abelian extension F/K, then $[x,K]|_F = \left(\frac{F/K}{\mathrm{il}(x)}\right)$, the Artin symbol of F/K.

Consider the special case $K = \mathbb{Q}$. Let $\mathbb{T}$ be the circle group $\{z \in \mathbb{C} \mid |z| = 1\}$, isomorphic image of the group $\mathbb{R}/\mathbb{Z}$ under the map $e^{2\pi i(-)} : \mathbb{R}/\mathbb{Z} \to \mathbb{T}$. Let U denote the subgroup of $\mathbb{T}$ of all roots of unity. For every $x \in \mathbb{Q}_A^*$ there is a commutative diagram

$$(2) \qquad \begin{array}{ccc} \mathbb{Q}/\mathbb{Z} & \longrightarrow & U \\ {\scriptstyle x^{-1}}\downarrow & & \downarrow{\scriptstyle [x,\mathbb{Q}]} \\ \mathbb{Q}/x^{-1}\mathbb{Z} & \longrightarrow & U \end{array}$$

where the upper horizontal arrow is the restriction of the map $e^{2\pi i(-)}$ to $\mathbb{Q}/\mathbb{Z}$, and the left vertical arrow can be obtained as follows. First, for $y \in \mathbb{Q}_A^*$, $y\mathbb{Z}$ is defined by $y\mathbb{Z} = il(y)$, i.e., $(y\mathbb{Z})_p = y_p\mathbb{Z}_p$ for all primes p, where y_p is the p-component of y. Next, observe that $\mathbb{Q}/\mathbb{Z}$ can be identified with the direct sum of $\mathbb{Q}_p/\mathbb{Z}_p$ for all p, and multiplication by y_p defines an isomorphism of $\mathbb{Q}_p/\mathbb{Z}_p$ onto $\mathbb{Q}_p/y_p\mathbb{Z}_p$. Combining these isomorphisms for all p, we obtain an isomorphism $y : \mathbb{Q}/\mathbb{Z} \to \mathbb{Q}/y\mathbb{Z}$. The left vertical arrow of (2) is obtained by taking x^{-1} as y. The lower horizontal arrow is obtained from the map $w \to e^{2\pi i(\alpha w)}$ with a unique $\alpha \in \mathbb{Q}$ such that $x^{-1}\mathbb{Z} = \alpha^{-1}\mathbb{Z}$ and $\alpha \cdot x_\infty > 0$, where x_∞ is the archimedean component of x. Finally, the right vertical arrow is simply the action of $[x,\mathbb{Q}]$ on the roots of unity. (Of course we know that $U \subset \mathbb{Q}_{ab}$).

The commutative diagram (2) is essentially equivalent with the classical formula

$$\zeta^{\left(\frac{\mathbb{Q}(\zeta)/\mathbb{Q}}{(m)}\right)} = \zeta^m$$

where $\zeta = e^{2\pi i/N}$, $m \in \mathbb{Z}$, $m > 0$, $(m,N) = 1$. Thus, in the special case $K = \mathbb{Q}$, the diagram (2) tells us something more than the exact sequence (1). It has also another classical fact

$$(3) \qquad \mathbb{Q}_{ab} = \mathbb{Q}(\{\zeta \mid \zeta \in U\}) = \mathbb{Q}(\{e^{2\pi i w} \mid w \in \mathbb{Q}\})$$

as its corollary.

Now one may ask a natural question : <u>Can one generalize</u> (2) <u>or</u> (3) <u>to an arbitrary algebraic number field</u> K ?

The purpose of Part I is to explain how this question is affirmatively answered for imaginary quadratic fields by means of complex multiplication of elliptic curves.

Shi-6

§2. The case of imaginary quadratic fields

Let K be any number field. One may attempt to set up a diagram of the type (2) for K as follows. Put for every rational prime number p

$$K_p = K \otimes_{\mathbb{Q}} \mathbb{Q}_p.$$

Let $a \subset K$ be a lattice, i.e. a $\mathbb{Z}$-module with the property $a \otimes_{\mathbb{Z}} \mathbb{Q} = K$. Then $a_p = a \otimes_{\mathbb{Z}} \mathbb{Z}_p$ is a $\mathbb{Z}_p$-lattice in K_p and it is easy to see that K/a is canonically isomorphic to the direct sum $\coprod K_p/a_p$. Let A be the adèles of $\mathbb{Q}$, then the adèle ring K_A of K is given by $K_A \cong K \otimes_{\mathbb{Q}} A$. Consequently, for every idèle $y \in K_A^*$ one can speak of its p-component y_p. We can again find a unique lattice ya in K such that $(ya)_p = y_p a_p$. For such a lattice we have a canonical isomorphism $K/ya \cong \coprod K_p/y_p a_p$. By the same idea as in the case $K = \mathbb{Q}$, we can define "multiplication with y" which gives an isomorphism $K/a \cong K/ya$. In this way we have "generalized" the left vertical arrow of (2) to the situation "K is a number field". The question becomes : what groups should be put instead of the U in (2) and what will the remaining three arrows have to be like ?

The answer can be given in case K is an <u>imaginary quadratic number field</u>, by taking as U the set of all points of finite order on an elliptic curve, whose endomorphism algebra is isomorphic to K. Let us now recall some elementary facts about endomorphisms of an elliptic curve. Let E be an elliptic curve; E is said to have <u>complex multiplication</u> if $\mathrm{End}(E) \neq \mathbb{Z}$. It is well-known that, in this case, $K_E = \mathrm{End}(E) \otimes_{\mathbb{Z}} \mathbb{Q}$ is an imaginary quadratic number field. Indeed, if $E = \mathbb{C}/\Lambda$, with Λ a lattice in $\mathbb{C}$, spanned by the primitive periods ω_1 and ω_2, then E has complex multiplication if and only if the quotient ω_1/ω_2 is imaginary quadratic; and in that case $K_E \cong \mathbb{Q}(\omega_1/\omega_2)$. Note also that $\mathbb{C}/\Lambda \cong \mathbb{C}/\omega_2^{-1}\Lambda$ with

$$\omega_2^{-1}\Lambda = \mathbb{Z} \oplus \mathbb{Z}\omega_1/\omega_2 \subset K,$$

where K is the subfield of $\mathbb{C}$ isomorphic to K_E.

Conversely, take any imaginary quadratic number field K, and let a be *any* lattice. Fix an embedding of K into $\mathbb{C}$. We have then an embedding $K/a \to \mathbb{C}/a$. We consider $\mathbb{C}/a$ as an elliptic curve E, or more precisely, we take an analytic isomorphism $\xi : \mathbb{C}/a \to E$, where E is a curve

$$Y^2 = 4X^3 - g_2X - g_3$$

in the two-dimensional projective space $\mathbb{P}^2$ over $\mathbb{C}$, with the coefficients g_2, g_3 determined by the lattice a in the usual manner, and

$$\xi(u) = (\wp(u), \wp'(u)),$$

where $\wp(z)$ is the Weierstrass $\wp$-function and $u \in \mathbb{C}/a$. Then E has complex multiplication and $\mathrm{End}(E) \otimes \mathbb{Q}$ is isomorphic to K. It is easy to see that $\xi(K/a) = E_o$ is just the set of points of finite order of E. With these notations and definitions we have the following theorem. (Note that, if σ is an arbitrary field automorphism of $\mathbb{C}$, then another elliptic curve E^σ is defined by the equation $Y^2 = 4X^3 - g_2^\sigma X - g_3^\sigma$. Obviously σ maps E_o onto $(E^\sigma)_o$).

THEOREM 1. *Let* K, $a \subset K$, ξ *and* E *be as above*. *Let* σ *be an automorphism of* $\mathbb{C}$ *and* $x \in K_A^*$ *an element with the property* $\sigma = [x,K]$ *on* K_{ab}. *Then, there is an analytic isomorphism* $\xi' : \mathbb{C}/x^{-1}a \to E^\sigma$ *such that the following diagram is commutative*

(4)
$$\begin{array}{ccc} K/a & \xrightarrow{\ \xi\ } & E_o \\ \downarrow{\scriptstyle x^{-1}} & & \downarrow{\scriptstyle \sigma} \\ K/x^{-1}a & \xrightarrow{\ \xi'\ } & E^\sigma{}_o \end{array}$$

For the proof of this theorem the reader is referred to [3], p.117.

We mention the following classical corollaries. Let a be a fractional ideal of K and let $j(\mathbb{C}/a) = j(E) = j(a)$ be the j-invariant of the curve

$E = \mathbb{C}/\mathfrak{a}$. Then $j(\mathfrak{a}) \in K_{ab}$, and we have the formula $j(\mathfrak{a})^{[x,K]} = j(x^{-1}\mathfrak{a})$, the extension $K(j(\mathfrak{a}))$ being the maximal unramified abelian extension of K. More classically, one can write the formula as

$$j(\mathfrak{a})^{\left(\frac{F/K}{\mathfrak{b}}\right)} = j(\mathfrak{b}^{-1}\mathfrak{a})$$

for any fractional ideal $\mathfrak{b}$ of K, where $F = K(j(\mathfrak{a}))$. It should be noted that the points of E_0 are not necessarily rational over K_{ab}. To settle this point more clearly, observe that Aut(E) is either of order 2,4 or 6, and define the function h(u) by :

$$h(u) = \begin{cases} \frac{g_2 g_3}{\Delta}.\wp(u) & \text{if Aut(E) is of order 2} \\ \frac{g_2^2}{\Delta}.\wp(u)^2 & \text{if Aut(E) is of order 4} \\ \frac{g_3}{\Delta}.\wp(u)^3 & \text{if Aut(E) is of order 6,} \end{cases}$$

where g_2, g_3 are as before the coefficients of the Weierstrass equation for E: $Y^2 = 4X^3 - g_2X - g_3$ and $\Delta = g_2^3 - 27g_3^2$. We then have the equality

$$K_{ab} = K(\{j(\mathfrak{a}), h(u) \mid u \in K/\mathfrak{a}\}),$$

which is an analogue of (3).

§3. The higher dimensional case

The results of §2 can be generalized, in a certain sense, from the case of an elliptic curve to particular types of abelian varieties. (Reminder : An abelian variety over $\mathbb{C}$ is a projective variety isomorphic to a complex torus as a complex manifold.) The fields K employed are so-called CM-fields. They are, by definition, totally imaginary quadratic extensions of totally real number fields. So, if $F/\mathbb{Q}$ is totally real of degree n and K/F totally imaginary of degree 2, then $[K : \mathbb{Q}] = 2n$. It is easy to see that the composite field of a finite number of CM-fields is a CM-field.

Every $\mathbb{Q}$-conjugate of a CM-field as well as the smallest Galois extension of $\mathbb{Q}$ containing a CM-field, are CM-fields. Let $\phi_1,\ldots,\phi_n$ be the n different embeddings of a given CM-field K, with $[K : \mathbb{Q}] = 2n$, into $\mathbb{C}$, whose restrictions to the maximal real subfield of K are different. One then obtains an $\mathbb{R}$-linear isomorphism $\phi : K_{\mathbb{R}} = K \otimes_{\mathbb{Q}} \mathbb{R} \to \mathbb{C}^n$ such that

$$\phi(x) = (x^{\phi_1},\ldots,x^{\phi_n})$$

for $x \in K$. Let $a \subset K$ be an arbitrary lattice. Then one can show that $\mathbb{C}^n/\phi(a)$ is a complex torus, and moreover has a structure of an <u>abelian variety</u> A of dimension n. Viewing this abelian variety as embedded into $\mathbb{P}^n$ and denoting by ξ the holomorphic isomorphism which identifies $\mathbb{C}^n/\phi(a)$ and $A \subset \mathbb{P}^n$, we might be tempted to expect the existence of a commutative diagram

$$(5)\qquad \begin{array}{ccccc}
K_{\mathbb{R}}/a & \xrightarrow{\phi} & \mathbb{C}^n/\phi(a) & \xrightarrow{\xi} & A \hookrightarrow \mathbb{P}^n \\
\cup\uparrow & & & & \Vert \\
K/a & & \xrightarrow{\xi\circ\phi} & & A \\
\downarrow x^{-1} & & & & \downarrow \sigma \\
K/x^{-1}a & & \xrightarrow{?} & & A^{\sigma}
\end{array}$$

for some $\sigma \in \mathrm{Aut}(\mathbb{C})$ with $\sigma|K_{ab} = [x,K]$, $x \in K_A^*$, just like before (diagram (4)). However, this is expecting too much, because if K' is a field isomorphic, but not equal, to K, then K_{ab} and K'_{ab}, though isomorphic, have different embeddings into $\mathbb{C}$; whereas the abelian varieties A attached to K and K' are the same, and the points of A have coordinates in $\mathbb{C}$. The commutativity of the diagram depends on the embedding of K_{ab} into $\mathbb{C}$! There is no way to specify such an embedding unless K is normal over $\mathbb{Q}$. The situation can be repaired as follows. We denote by Φ the direct sum of the n isomorphisms $\phi_1,\ldots,\phi_n$ of the given CM-field K into $\mathbb{C}$. So for $x \in K$ we have

$$\Phi(x) = \begin{pmatrix} x^{\phi_1} & & & \\ & x^{\phi_2} & & \\ & & \ddots & \\ & & & x^{\phi_n} \end{pmatrix}$$

Define a subfield K' of $\mathbb{C}$ by $K' = \mathbb{Q}(\{\mathrm{tr}\Phi(x) \mid x \in K\})$. Then K' is a CM-field. This may be shown by using the criterion saying that a field K is a CM-field if and only if the following two conditions are satisfied:

(i) If ρ denotes the operation of taking complex conjugates, then ρ induces a non-trivial automorphism of K,

(ii) $\rho\tau = \tau\rho$ for every isomorphism τ of K into C.

Now observe that, for any number fields F and F', every $(F \otimes_{\mathbb{Q}} F')$-module V can produce a representation $F \hookrightarrow \mathrm{End}(V,F')$ of the field F in a natural manner. If the rank of V over F' is finite, say m, then one obtains a representation of F by matrices of size m over F.

PROPOSITION. <u>With</u> K' <u>defined as above for a</u> CM-<u>field</u> K <u>and</u> Φ, <u>there exists a</u> $K \otimes_{\mathbb{Q}} K'$-<u>module</u> V <u>such that the representation</u> $K \hookrightarrow \mathrm{End}(V,K')$ <u>is equivalent to</u> Φ <u>over</u> $\mathbb{C}$. V <u>is, up to isomorphism, unique.</u>

For such a V, one obtains a representation $\Phi' : K' \hookrightarrow \mathrm{End}(V,K)$. Thus we are able to produce a new couple (K',Φ') from a given couple (K,Φ). Now, it is easy to see that for all $y \in K'$, $\det \Phi'(y) \in K$; hence the map $\det \Phi' : K'^* \to K^*$ can be extended to a continuous homomorphism of $K'^*_A \to K^*_A$ of the idèles of K' and K. With these notations and definitions we have

THEOREM 2. <u>For</u> $x \in K'^*_A$, $\sigma \in \mathrm{Aut}(C)$ <u>such that</u> $\sigma = [x,K']$ <u>on</u> K'_{ab}, <u>there exists a holomorphic isomorphism</u> $\xi' : \mathbb{C}^n/\phi(\det\Phi'(x)^{-1}a) \to A^\sigma$ <u>making the</u>

diagram

$$\begin{array}{ccc} K/\mathfrak{a} & \xrightarrow{\ \xi\circ\phi\ } & A \\ {\scriptstyle \det\Phi'(x)^{-1}}\downarrow & & \downarrow{\scriptstyle\sigma} \\ K/\det\Phi'(x)^{-1}\mathfrak{a} & \xrightarrow{\ \xi'\circ\phi\ } & A \end{array}$$

commutative.

PART II. ELLIPTIC CURVES WITH COMPLEX MULTIPLICATION UNIFORMIZED BY MODULAR FUNCTIONS

Let us first discuss the zeta-function of a "factor" A of the Jacobian of a modular function field of an arbitrary level. After stating the results about this in the general case, we shall consider the question "when does such an A have complex multiplication ?", and then connect our study with Deuring's results on the zeta-function of an elliptic curve with complex multiplication.

Let S_N denote the vector space of all cusp forms of weight 2 with respect to the group

$$\Gamma_1(N) = \{\left[\begin{smallmatrix} a & b \\ c & d \end{smallmatrix}\right] \in SL_2(\mathbb{Z}) \mid c \equiv 0,\ a \equiv d \equiv 1 \pmod N\}.$$

Let J_N denote the jacobian variety (defined over $\mathbb{Q}$) of the standard $\mathbb{Q}$-rational model of the curve $H^*/\Gamma_1(N)$, where H^* denotes the union of the upper half plane H and the cusps.
For an arbitrary curve or an abelian variety W over $\mathbb{C}$, let D(W) denote the vector space of all holomorphic 1-forms on W. Then we have a canonical isomorphism

$$\mu : S_N \to D(J_N)$$

obtained by identifying S_N with $D(H^*/\Gamma_1(N))$, and then $D(H^*/\Gamma_1(N))$ with $D(J_N)$.

Take a common eigen-function f in S_N of the Hecke operators T_n for all n. Let $f(z) = \Sigma_{n=1}^{\infty} a_n e^{2\pi inz}$ be the Fourier expansion of f. We normalize f so that $a_1 = 1$; then $f|T_n = a_n f$. With a fixed f, consider the field $F = \mathbf{Q}(a_1, a_2, \ldots)$ generated by the coefficients a_n. It can be shown that F is an algebraic number field of finite degree. Note that F depends on f.

THEOREM 3. For a normalized eigen-function f as above, there exists a pair (A, θ) formed by the objects satisfying the following conditions.

(i) A is a quotient of J_N by an abelian subvariety rational over $\mathbf{Q}$. (Denote by ν the natural map of J_N onto A.)

(ii) θ is an injective homomorphism of F into $\mathrm{End}(A) \otimes \mathbf{Q}$ such that $\nu \circ \xi_n = \theta(a_n) \circ \nu$ for all n, where ξ_n is the element of $\mathrm{End}(J_N)$ corresponding to T_n.

(iii) $\dim(A) = [F : Q]$.

Such a pair (A, θ) is unique up to isomorphism under the conditions (i), (ii). Moreover, let I denote the set of all isomorphisms of F into $\mathbf{C}$. Then for every $\sigma \in I$, there exists an element f_σ of S_N such that $f_\sigma|T_n = a_n^\sigma$ for all n, and $f_\sigma(z) = \Sigma_{n=1}^{\infty} a_n^\sigma e^{2\pi inz}$. With these f_σ, one has

(iv) $\delta\nu(D(A)) = \mu(\Sigma_{\sigma\in I}\mathbf{C}f_\sigma)$, where $\delta\nu$ is the "pull back" : $D(A) \to D(J_N)$ attached to ν.

THEOREM 4. The notation being as above, suppose that the vector subspace $\Sigma_{\sigma\in I}\mathbf{C}f_\sigma$ is stable under the map $h(z) \mapsto z^{-2}h(-1/Nz)$. Then the zeta-function of A over $\mathbf{Q}$ is almost equal to $\Pi_{\sigma\in I}(\Sigma_{n=1}^{\infty} a_n n^{-s})$. (The expression "almost equal" means that the product $\Pi_{\sigma\in I}(\Sigma_{n=1}^{\infty} a_n n^{-s})$ has the right p-Euler factor for all primes p not dividing N.)

As a supplementary result, we have

PROPOSITION 1. *Let* $m = [F : \mathbb{Q}]$, *and let* P *be the* $\mathbb{Z}$*-submodule of* $\mathbb{C}^m$ *generated over* $\mathbb{Z}$ *by the vectors*

$$v(\gamma) = \left(\int_w^{\gamma(w)} f_\sigma(z)dz \right)_{\sigma \in I}$$

for all $\gamma \in \Gamma_1(N)$. *(Note that* $v(\gamma)$ *is independent of the choice of* w *on* H.) *Then* A *is isomorphic to the complex torus* $\mathbb{C}^m/P$.

Theorems 3 and 4 are, in essence, equivalent to the results of [3, §7.5], where we considered abelian *subvarieties* of J_N instead of *quotients* of J_N. The proofs of the theorems and Prop.1, together with some other results, will appear in [6].

We now turn to the case of elliptic curves with complex multiplication.

THEOREM 5. (Deuring [1]). *Let* E *be an elliptic curve with complex multiplication, defined over an algebraic number field* k, *and* K *the imaginary quadratic field isomorphic to* $\mathrm{End}(E) \otimes \mathbb{Q}$. *Let* $Z(s, E/k)$ *denote the zeta-function of* E *over* k, *and* $L(s,\lambda)$ *the* L-*function* $\Sigma_{\mathfrak{a}}\lambda(\mathfrak{a})N(\mathfrak{a})^{-s}$ *with a Grössen-character* λ *of* k. (Of course s is a complex variable, $N(\mathfrak{a})$ the norm of an ideal $\mathfrak{a}$ of k; $\mathfrak{a}$ runs over all integral ideals of k prime to the conductor of λ.) *Then, if* $K \subset k$, *we have* $Z(s,E/k) = L(s,\lambda)L(s,\bar{\lambda})$ *with a primitive Grössen-character* λ *of* k; *if* $K \not\subset k$, *we have* $Z(s,E/k) = L(s,\lambda)$ *with a primitive Grössen-character* λ *of the composite field* Kk.

We are going to connect this result with cusp forms of weight 2. Let K be an imaginary quadratic field with discriminant $-D$, and λ a Grössen-character of K with conductor $\mathfrak{f}$. We consider λ as a character of the ideal-group modulo $\mathfrak{f}$, and not as a character of the idèle group. Furthermore we restrict our discussion to the characters λ such that $\lambda((\alpha)) = \alpha^m$ for $\alpha \in K$, $\alpha \equiv 1 \pmod{\mathfrak{f}}$, with a positive integer m. Now define a function $f_\lambda(z)$ on H by $f_\lambda(z) = \Sigma_{\mathfrak{a}}\lambda(\mathfrak{a})e^{2\pi i N(\mathfrak{a})}$ (i.e., f_λ is the Mellin transform of the L-function $L(s,\lambda) = \Sigma_{\mathfrak{a}}\lambda(\mathfrak{a})N(\mathfrak{a})^{-s}$). Then it can be shown that f_λ is a cusp form of weight m + 1 with respect to $\Gamma_1(N)$, where $N = D.N(\mathfrak{f})$. (This is due to Hecke [2]; see also [4].) In particular, assume m = 1. Then $f_\lambda \in S_N$. Moreover, the existence of Euler product for $L(s,\lambda)$ implies that f_λ is a normalized common eigen-function of all the Hecke operators. Therefore, applying Theorem 3 to f_λ, we can let correspond to λ a pair (A,θ). In this situation, we have

THEOREM 6. There is an elliptic curve E such that A is isogenous to $E \times \ldots \times E$, and $\mathrm{End}(E) \otimes \mathbb{Q} \cong K$.

For the proof, see [4]. This means especially that every elliptic curve with complex multiplication is a "factor" of J_N for some N (of course over $\mathbb{C}$ and up to isogeny). A natural question arises whether or not an abelian variety of CM-type of higher dimension as considered in Part I, §3 can occur as a factor of J_N. The answer is negative; in fact, one has :

THEOREM 7. (Casselman). Let A be a simple abelian subvariety of J_N of CM-type. Then A is an elliptic curve. (We say that an abelian variety A is of CM-type, if it is obtained in the manner described in Part I, §3.)

More precisely, if a new form (in the sense of Atkin-Lehner-Miyake) f(z) of weight 2 in S_N gives rise to (A,θ), and if A has an abelian subvariety of CM-type, then A is isogenous to $E \times \ldots \times E$ with an elliptic curve E with complex multiplication; moreover, $f = f_\lambda$ with a primitive Grössen-character λ of the imaginary quadratic field K such that $\mathrm{End}(E) \otimes \mathbb{Q} \cong K$. For the proof, see [5,pp 137-138].

In particular, consider an elliptic curve rational over $\mathbb{Q}$ with complex multiplication, and let $\mathrm{End}(E) \otimes \mathbb{Q} \cong K$ (this happens if and only if K has class number 1). Then, by Theorem 5, the zeta-function of E over $\mathbb{Q}$ is $L(s,\lambda)$ with a Grössen-character λ of K. Further, by Theorem 3, we can attach (A,θ) to f_λ. In brief, we have

$$E \xmapsto{\text{Th.5}} \lambda \longmapsto f_\lambda \xmapsto{\text{Th.3}} A.$$

By Theorem 6, we know that $\mathrm{End}(A) \otimes \mathbb{Q} \cong K$. (Note that $\dim (A) = [F : \mathbb{Q}] = 1$, since the Fourier coefficients of f_λ belong to $\mathbb{Q}$.) Therefore A and E are isogenous over $\mathbb{C}$, but actually we have

THEOREM 8. E _is isogenous to_ A _over_ $\mathbb{Q}$.

This is because the zeta-functions of A and E are almost equal. For details, see [4].

Remark. (i) Deuring [1, IV] showed that E has good reduction modulo p if and only if $p \nmid D.N(\mathfrak{f})$, where $\mathfrak{f}$ is the conductor of λ.

(ii) By Prop.1, we know that the quotient

$$\int_w^{\alpha(w)} f_\lambda(z)dz \Big/ \int_w^{\beta(w)} f_\lambda(z)dz$$

belongs to K for any two elements α and β of $\Gamma_1(N)$.

PART III. COMPLEX MULTIPLICATION IN TERMS OF MODULAR FUNCTIONS

Let F_N denote the set of all modular functions $f(z)$ with respecto to $\Gamma(N)$ such that $f(z) = \Sigma_{n \geq n_0} c_n e^{2\pi inz/N}$ with $c_n \in \mathbb{Q}(\zeta_N)$, where $\zeta_N = e^{2\pi i/N}$, and $\Gamma(N)$ is the principal congruence subgroup of $SL_2(\mathbb{Z})$ of level N. It can easily be shown that F_N forms a field with the following properties :

Shi-16

(i) The composite $F_N.\mathbb{C}$ is the field of all modular functions with respect to $\Gamma(N)$.

(ii) F_N is linearly disjoint with $\mathbb{C}$ over $\mathbb{Q}(\zeta_N)$.

(iii) $F_1 = \mathbb{Q}(j)$ with the well-known j (i.e. the modular invariant).

(iv) F_N is a Galois extension of F_1 with Galois group $GL_2(\mathbb{Z}/N\mathbb{Z})/\{\pm 1\}$.

The action of $GL_2(\mathbb{Z}/N\mathbb{Z})/\{\pm 1\}$ on F_N can be explicitly described as follows. For every $\xi \in SL_2(\mathbb{Z})$ and $h \in F_N$, we take $h^{\xi \bmod N} = h \circ \xi$, via the operation of ξ on the upper half plane H. For every $\alpha \in GL_2(\mathbb{Z}/N\mathbb{Z})$, we have $\zeta_N^{\alpha} = \zeta_N^{\det(\alpha)}$, and for the representatives of $GL_2(\mathbb{Z}/N\mathbb{Z})/\{\pm 1\}$ modulo $SL_2(\mathbb{Z}/N\mathbb{Z})/\{\pm 1\}$ of the form $\alpha = \pm \begin{bmatrix} 1 & 0 \\ 0 & k \end{bmatrix} \bmod N$, we have

$$(\Sigma c_n e^{2\pi i n z/N})^{\alpha} = \Sigma c_n^{\sigma_k} e^{2\pi i n z/N}$$

where σ_k is the element of $Gal(\mathbb{Q}(\zeta_N)/\mathbb{Q})$ such that $\zeta_N^{\sigma_k} = \zeta_N^k$.
Now let us take the union $F = \cup_{n=1}^{\infty} F_N$, which is an infinite Galois extension of F_1, with

$$Gal(F/F_1) \cong (\Pi_p GL_2(\mathbb{Z}_p))/\{\pm 1\},$$

where the product extends over all rational primes p. This can easily be seen from the above structure of $Gal(F_N/F_1)$.
Next consider <u>all</u> automorphisms of F, which are not necessarily identity on F_1. Of course they form a group, which we denote by Aut(F). Moreover, we can make Aut(F) a locally compact topological group by taking

$$\{Gal(F/F_N) \mid N = 1,2,\ldots\}$$

as a base of neighborhoods of the identity element.

THEOREM 9. <u>There exists an exact sequence</u>

$$1 \longrightarrow \mathbb{Q}^* G_{\infty +} \longrightarrow G_{A+} \xrightarrow{\tau} Aut(F) \longrightarrow 1.$$

Here $G_A = GL_2(A) = G_{finite}G_\infty$ (A = the adèles of Q), $G_\infty = GL_2(\mathbf{R})$, $G_{A+} = G_{finite}G_{\infty+}$ with $G_{\infty+}$ the identity component of $GL_2(\mathbf{R})$, and Q^* is the group of scalar matrices in $G_Q = GL_2(Q)$ embedded in G_A. The map τ is continuous, and $Q^*G_{\infty+}$ is closed in G_{A+}.

The reader's attention may be drawn to the fact that the above exact sequence is analogous to that of class field theory (Part I, §1, (1)).

The action of G_{A+} on F, via τ, can be described as follows.

(i) <u>Action of $G_\mathbb{Q}$ on F</u> : Let $h \in F$ and $\alpha \in G_\mathbb{Q}$, $\det(\alpha) > 0$. Then $h^{\tau(\alpha)} = h \circ \alpha$.

(ii) <u>Action on the series $\sum_{n>n_o} c_n e^{2\pi inz/N}$</u> : For $c_n \in \mathbb{Q}_{ab}$ and $x \in \mathbb{Q}_A^*$, $x_\infty > 0$, one has

$$\left(\sum_{n>n_o} c_n e^{2\pi inz/N}\right)^{\tau\left(\left[\begin{smallmatrix}1 & 0\\ 0 & x\end{smallmatrix}\right]\right)} = \sum_{n>n_o} c_n^{[x^{-1},\mathbb{Q}]}\, e^{2\pi inz/N},$$

where $[x^{-1},\mathbb{Q}]$ is the action of x^{-1} on $\mathbb{Q}_{ab}$ (see Part I, §1).

(iii) <u>Action on $\mathbb{Q}_{ab}$</u> : For $\xi \in G_{A+}$ and $c \in \mathbb{Q}_{ab}$, one has $c^{\tau(\xi)} = c^{[\det(\xi)^{-1},\mathbb{Q}]}$. In other words, the following diagram is commutative.

$$\begin{array}{ccc} G_{A+} & \xrightarrow{\ \tau\ } & \mathrm{Aut}(F) \\ \Big\downarrow \det(\ )^{-1} & & \Big\downarrow \\ \mathbb{Q}_A^* & \longrightarrow & \mathrm{Gal}(\mathbb{Q}_{ab}/\mathbb{Q}) \end{array}$$

We can now formulate complex multiplication in terms of modular functions by means of the map τ. Take any imaginary quadratic number field K embedded in $\mathbf{C}$, and let $g : K \to M_2(\mathbb{Q})$ be a representation of K by rational matrices of degree 2. The image of K^* in $GL_2(\mathbb{Q})$, $g(K^*)$, is a set of mutually commutative elliptic transformations of the upper half plane H, so that there is a unique common fixed point, say z_g, of the elements of $g(K^*)$. Changing g for the map $a \longmapsto g(\bar{a})$ if necessary, we may assume that

$$g(a)\begin{bmatrix} z_g \\ 1 \end{bmatrix} = a\begin{bmatrix} z_g \\ 1 \end{bmatrix}$$

for all $a \in K$. We call g normalized if the last condition is satisfied. We then extend g to a continous map of K_A^* into G_{A+}, and denote it again by g. With this setting we have

THEOREM 10. For all $h \in F$, we have $h(z_g) \in K_{ab}$, and

$$h(z_g)^{[x,K]} = h^{\tau(g(x)^{-1})}(z_g)$$

for all $x \in K^*$.

This is similar to the fact that the canonical map $K_A^* \to \mathrm{Gal}(K_{ab}/K)$ is defined locally by assigning the Frobenius map to the prime element. We may symbolically say that Theorems 9 and 10 represent a sort of two-dimensional (in the Kroneckerian sense) class field theory.

From the above theorem, we can easily derive

THEOREM 11. Let K, g and z_g be as above. Further, for a fixed $h \in F$, other than constants, put $S_h = \{x \in G_{A+} | h^{\tau(x)} = h\}$. Suppose that F is Galois over $\mathbf{Q}(h)$. (This is so if and only if $\mathbf{Q}(h) = \{f \in F | f^{\tau(x)} = f$ for all $x \in S_h\}$.) Then $K(h(z_g))$ is the subfield of K_{ab} corresponding to the subgroup $\{x \in K_A^* | g(x) \in S_h\}$ of K_A^*.

The condition "F is Galois over $\mathbf{Q}(h)$" is satisfied whenever the function $j(z)$ is contained in $\mathbf{Q}(h)$. This is so, for example, if h is one of the classical functions $\gamma_2, \gamma_3, f, f_1, f_2$ defined by

$$\gamma_2(z) = (2\pi)^{-4} 12 g_2 \eta^{-8}, \quad \gamma_3(z) = \sqrt{27}\,(2\pi)^{-6} g_3 \eta^{-12},$$

$$f(z) = e^{-\pi i/24}\eta((z+1)/2)/\eta(z),$$

$$f_1(z) = \eta(z/2)/\eta(z), \quad f_2(z) = \sqrt{2}\,\eta(2z)/\eta(z),$$

where $$\eta(z) = e^{\pi i z/12}\prod_{n=1}^{\infty}(1 - e^{2\pi i n z}).$$

In fact, we have the following relations :

$$j = \gamma_2^3 = \gamma_3^2 + 12^3, \quad \gamma_2 = (f^{24} - 16)/f^8.$$

For details, the reader is referred to Weber [7].

Therefore, to determine the fields generated by the values of these functions at the special points z_g, we only have to determine

$$S_h \text{ and } \{x \in K_A^* | g(x) \in S_h\},$$

which must be an easy (if tedious) exercise for anybody who has a little experience on a similar type of question.

REFERENCES

[1] M. DEURING, Die Zetafunktion einer algebraischen Kurve vom Geslechte Eins, I,II,III,IV, Nachr. Akad. Wiss. Göttingen, (1953) 85-94, (1955) 13-42, (1956) 37-76, (1957) 55-80.

[2] E. HECKE, Zur Theorie der elliptischen Modulfunktionen, Math. Ann., 97 (1926), 210-242 (= Math. Werke, 428-460).

[3] G. SHIMURA, Introduction to the arithmetic theory of automorphic functions, Publ. Math. Soc. Japan, N°.11, Iwanami Shoten and Princeton Univ. Press, 1971.

[4] G. SHIMURA, On elliptic curves with complex multiplication as factors of the jacobians of modular function fields, Nagoya Math. J. 43 (1971), 199-208.

[5] G. SHIMURA, Class fields over real quadratic fields and Hecke operators, Ann. of Math. 95 (1972), 130-190.

[6] G. SHIMURA, On the factors of the jacobian variety of a modular function field, to appear.

[7] H. WEBER, Lerhbuch der Algebra III.

MODULAR FORMS OF HALF INTEGRAL WEIGHT

By Goro Shimura

International Summer School on Modular Functions
Antwerp 1972

CONTENTS

MODULAR FORMS OF HALF INTEGRAL WEIGHT

§1. Definitions and elementary properties

The forms to be discussed are those with the automorphic factor $(cz + d)^{k/2}$ with a positive odd integer k. The theta function

$$\theta(z) = \sum_{n=-\infty}^{\infty} e^{2\pi i n^2 z}$$

and the Dedekind eta function

$$\eta(z) = e^{\pi i z/12} \prod_{n=1}^{\infty} (1 - e^{2\pi i n z})$$

are classical examples of such forms. (For some practical reasons, we take $e^{2\pi i n^2 z}$ instead of the usual $e^{\pi i n^2 z}$ in the definition of θ.) In fact, the function θ satisfies

(1.1) $$\theta(\gamma(z)) = j(\gamma, z)\, \theta(z) \quad \text{for all} \quad \gamma \in \Gamma_0(4)$$

with

(1.2) $$j(\begin{bmatrix} a & b \\ c & d \end{bmatrix}, z) = (\tfrac{c}{d})\, \varepsilon_d^{-1}\, (cz + d)^{1/2}.$$

Here and henceforth, z is the standard variable on the upper half plane

$$H = \{z \in \mathbf{C} \mid \mathrm{Im}(z) > 0\};$$

$\Gamma_0(4)$ is defined by (1.6) below; the quadratic residue symbol $(\frac{c}{d})$ and the constant ε_d are defined so that

(1.3)
$$(\tfrac{c}{d}) = \begin{cases} -(\frac{c}{|d|}) & \text{if } c < 0, \ d < 0, \\ (\frac{c}{|d|}) & \text{otherwise}, \end{cases}$$

$$\varepsilon_d = \begin{cases} 1 & \text{if } d \equiv 1 \pmod 4, \\ i = \sqrt{-1} & \text{if } d \equiv 3 \pmod 4. \end{cases}$$

Shi-24

As for the square root $(cz + d)^{1/2}$, we always choose $w^{1/2}$ so that $-\pi/2 < \arg(w^{1/2}) \leq \pi/2$, and put $w^{k/2} = (w^{1/2})^k$ for $k \in \mathbb{Z}$. Note that

$$(1.4) \qquad j(-1, z) = 1,$$

$$(1.5) \qquad j(\begin{bmatrix} a & b \\ c & d \end{bmatrix}, z)^2 = (\tfrac{-1}{d})\ (cz + d) \qquad (\begin{bmatrix} a & b \\ c & d \end{bmatrix} \in \Gamma_0(4)).$$

An obvious difficulty in dealing with such automorphic factors is the multivaluedness of $(cz + d)^{k/2}$. To make our discussion smooth in this respect, or to pretend it, we introduce a group G consisting of all couples $(\alpha, \phi(z))$ formed by $\alpha = \begin{bmatrix} a & b \\ c & d \end{bmatrix} \in GL_2(\mathbb{R})$ with $\det(\alpha) > 0$ and a holomorphic function $\phi(z)$ on H such that

$$\phi(z)^2 = t\cdot\det(\alpha)^{-1/2}\ (cz + d)$$

with $t \in \mathbb{C}$, $|t| = 1$. The group law of G is defined by $(\alpha, \phi(z)) (\beta, \Psi(z)) = (\alpha\beta, \phi(\beta(z))\ \Psi(z))$. We define the action of (α, ϕ) on H to be the same as that of α, and put

$$G_1 = \{(\alpha, \phi) \in G \mid \det(\alpha) = 1\}.$$

(Actually this is the simplest case of the metaplectic group introduced in Weil [12].) For an integer k, a function f(z) on H, and an element $\zeta = (\alpha, \phi)$ of G, we put

$$f|[\zeta]_k = f(\alpha(z))\ \phi(z)^{-k}.$$

Then $f|[\zeta\eta]_k = (f|[\zeta]_k) \mid [\eta]_k$. In the following treatment, k will always be <u>odd</u>.

Although one can develop the theory of automorphic forms of weight k/2 for a rather wide class of discrete subgroups of G_1, let us consider here only the case of "congruence subgroups". Let N be a multiple of 4, and let

$$(1.6)\qquad \begin{aligned} \Gamma_0(N) &= \{[\begin{smallmatrix} a & b \\ c & d \end{smallmatrix}] \in SL_2(\mathbb{Z}) \mid c \equiv 0 \pmod N\}, \\ \Gamma_1(N) &= \{[\begin{smallmatrix} a & b \\ c & d \end{smallmatrix}] \in \Gamma_0(N) \mid a \equiv d \equiv 1 \pmod N\}. \end{aligned}$$

Then we can define subgroups $\Delta_0(N)$ and $\Delta_1(N)$ of G_1 by

$$(1.7)\qquad \begin{aligned} \Delta_0(N) &= \{(\gamma, j(\gamma, z)) \mid \gamma \in \Gamma_0(N)\}, \\ \Delta_1(N) &= \{(\gamma, j(\gamma, z)) \mid \gamma \in \Gamma_1(N)\}. \end{aligned}$$

(Note that $\gamma \mapsto (\gamma, j(\gamma, z))$ defines an injection of $\Gamma_0(4)$ into G_1.) Let k be a positive odd integer, and Δ a subgroup of $\Delta_0(4)$ of finite index. Then a holomorphic function $f(z)$ on H is called an <u>integral modular form of weight</u> $k/2$ <u>with respect to</u> Δ, if $f|[\delta]_k = f$ for all $\delta \in \Delta$, and if it is "holomorphic" at each cusp s of $SL_2(\mathbb{Z})$ in the following sense. Let s be a cusp, and take $\rho \in G_1$ and a positive real number h so that $\rho(\infty) = s$, and $\rho(\pm[\begin{smallmatrix} 1 & h \\ 0 & 1 \end{smallmatrix}], t)\rho^{-1}$ generates the free cyclic part of $\{\eta \in \Delta | \eta(s) = s\}$. Here $t \in \mathbf{C}$, $|t| = 1$; it can be shown that t depends only on the Δ-equivalence class of s. Put $t^k = e^{2\pi ir}$ with $0 \leqslant r < 1$. Then we call f <u>holomorphic</u> at s if f has an expansion of the form

$$f|[\rho]_k = \textstyle\sum_{n \geqslant 0} c_n e^{2\pi i((n+r)z/h)}$$

with $c_n \in \mathbf{C}$. The cusp s is said to be k-<u>regular</u> or k-<u>irregular with respect to</u> Δ, according as $t^k = 1$ or not. We call f a <u>cusp form</u> if $c_0 = 0$ at every k-regular cusp s. Note that f has automatically "zero" of some fractional order at each k-irregular cusp. We denote by $M_k(\Delta)$ resp. $S_k(\Delta)$ the vector space of all integral modular forms resp. all cusp forms of weight $k/2$ with respect to Δ.

For example, $\Delta_0(4)$ has three inequivalent cusps represented by 0, ∞, $1/2$. The first two are k-regular for every odd k; the cusp $1/2$ is k-irregular with $t^k = e^{k\pi i/2}$.

Now one can define Eisenstein series of weight $k/2$ with respect to Δ in a natural manner, and show that $M_k(\Delta)$ is the direct sum of $S_k(\Delta)$ and the space of Eisenstein series. Moreover, $\dim(M_k(\Delta)) - \dim(S_k(\Delta))$ is the number of k-regular cusps if $k \geq 5$. For these and other related results, the reader is referred to Petersson [4], [5] and Maass [3].

If f is a cusp form, the usual argument yields the estimate of the Fourier coefficients:

$$c_n = O(n^{k/4}).$$

A more general result applicable to any element of $M_k(\Delta)$ and any k is given in Petersson [6].

The dimension of $M_k(\Delta)$ or $S_k(\Delta)$ can easily be computed by means of the Riemann-Roch theorem, if $k \geq 5$. For example,

$$\dim M_k(\Delta_0(4)) = [\tfrac{k}{4}] + 1 \qquad (k > 0),$$

$$\dim S_k(\Delta_0(4)) = \begin{cases} 0 & (k < 9), \\ [\tfrac{k}{4}] - 1 & (k \geq 9). \end{cases}$$

Here $[x]$ denotes the largest integer $\leq x$.

Let χ be a character of $(\mathbf{Z}/N\mathbf{Z})^*$. Then we denote by $M_k(N, \chi)$ (resp. $S_k(N, \chi)$) the subspace of $M_k(\Delta_1(N))$ (resp. $S_k(\Delta_1(N))$ formed by the elements f satisfying

$$f|[\delta]_k = \chi(d)f \quad \text{for all} \quad \delta = (\gamma, j(\gamma, z)), \quad \gamma = \begin{bmatrix} a & b \\ c & d \end{bmatrix} \in \Gamma_0(N).$$

Since $j(-1, z) = 1$, $M_k(N, \chi)$ can be non-trivial only if $\chi(-1) = 1$, so that we shall always assume $\chi(-1) = 1$ in the following treatment.

The functions θ and η are actually special cases of theta series

$$\theta(z, A, N, h) = \sum \exp(2\pi iz \cdot {}^t mAm/N).$$

Here A is a positive definite symmetric matrix of an odd size k with integral coefficients; the sum is extended over all $m \in \mathbf{Z}^k$ under the condition $m \equiv h \bmod N\mathbf{Z}^k$; we assume $h \in \mathbf{Z}^k$, $Ah \in N\mathbf{Z}^k$, and NA^{-1} has integral coefficients. Such a theta series is an element of $M_k(\Delta_1(4N))$. (For details, see e.g. [8, §2].) As a special case, take a primitive character Ψ of $(\mathbf{Z}/r\mathbf{Z})^*$, and put

$$(1.8) \qquad h_\Psi = (1/2) \sum_{n=-\infty}^{\infty} \Psi(n)n^\nu e^{2\pi i n^2 z},$$

where ν is either 0 or 1, and defined by $\Psi(-1) = (-1)^\nu$. Then it can be shown that $h_\Psi \in M_{2\nu+1}(4r^2, \Psi_1)$ with $\Psi_1(d) = \Psi(d)\,(\frac{-1}{d})^\nu$. Especially $2h_\Psi = \theta$ if Ψ is the trivial character. Further one has $h_\Psi(z) = \eta(24z)$ with $\Psi(d) = (\frac{3}{d})$. It should be noted that the Mellin transform of h_Ψ is the Dirichlet L-function

$$L(2s - \nu, \Psi) = \sum_{m=1}^{\infty} \Psi(m)m^{\nu-2s}$$

if $\Psi \neq 1$; it is $\zeta(2s)$ if $\Psi = 1$.

2. Hecke operators and the main theorem

Hecke realized the difficulty in developing the theory of his operators for the forms of half integral weight, and discussed only a very special case in his last paper [1]. The problem was later taken up by Wohlfahrt [14]. He defined a Hecke operator of degree p^2 (so to speak) for each prime p not dividing the level, and obtained a certain multiplicative relation between the Fourier coefficients of an eigen-function. Let us first define these operators in a somewhat more transparent and general form.

Fix a positive multiple N of 4, and for each positive integer m, put $\Delta = \Delta_0(N)$, $\alpha = \begin{bmatrix} 1 & 0 \\ 0 & m \end{bmatrix}$, $\zeta = (\alpha, m^{1/4})$, and consider a disjoint coset decomposition $\Delta\zeta\Delta = \cup_\nu \Delta\zeta_\nu$. Then we define an operator T(m) on $M_k(N, \chi)$ by

$$f|T(m) = m^{k/4-1} \sum_\nu \chi(a_\nu) f|[\zeta_\nu]_k,$$

where $\zeta_\nu = ([\begin{smallmatrix} a_\nu & * \\ * & * \end{smallmatrix}], *)$. It can easily be shown that $T(m)$ maps $M_k(N, \chi)$ and $S_k(N, \chi)$ into themselves. Now we have

$$T(m) = 0 \text{ \underline{if} } (m, N) = 1 \text{ \underline{and} } m \text{ \underline{is not a square}}.$$

This can be derived from the relation

$$\zeta\gamma^*\zeta^{-1} = (\alpha\gamma\alpha^{-1})^* \ (1, (\tfrac{m}{d}))$$

$$\text{for} \quad \gamma = [\begin{smallmatrix} a & b \\ c & d \end{smallmatrix}] \in \Gamma_0(4) \cap \alpha^{-1}\Gamma_0(4)\alpha,$$

where we put $\gamma^* = (\gamma, j(\gamma, z))$. Thus we may have non-trivial operators only for square m or for $(m, N) \neq 1$. This is discouraging, but still we have the following rather amusing result:

THEOREM 1: *Let* p *be a prime, and* f *an element of* $M_k(N, \chi)$. *Put*

$$f(z) = \textstyle\sum_{n=0}^{\infty} a(n)e^{2\pi inz},$$

$$(f|T(p^2))(z) = \textstyle\sum_{n=0}^{\infty} b(n)e^{2\pi inz}.$$

Then

$$b(n) = a(p^2n) + \chi_1(p)(\tfrac{n}{p})\ p^{\lambda-1}a(n) + \chi(p^2)p^{k-2}a(n/p^2),$$

where $\lambda = (k-1)/2$, $\chi_1(m) = \chi(m)(\frac{-1}{m})^\lambda$; *we put* $a(n/p^2) = 0$ *if* p^2 *does not divide* n, *and* $\chi(m) = \chi_1(m) = 0$ *if* $(m, N) \neq 1$.

This can easily be verified by writing explicitly the representatives ζ_ν of $\Gamma_0\zeta\Gamma_0$, and checking their effect on the Fourier coefficients. It is noteworthy that the computation involves the Gauss sum $\sum(\frac{m}{p})e^{2\pi im/p}$.

THEOREM 2: *Let* f *be as above. Suppose that* f *is a common eigen-function of* $T(p^2)$ *for all primes* p, *and put* $f|T(p^2) = \omega_p f$. *Let* t *be a positive integer which has no square factor prime to* N, *other than* 1. *Then one has, formally,*

$$\sum_{n=1}^{\infty} a(tn^2)n^{-s} = a(t)\ \Pi_p[1 - \chi_1(p)(\tfrac{t}{p})p^{\lambda-1-s}]\ [1 - \omega_p p^{-s} + \chi(p)^2 p^{k-2-2s}]^{-1},$$

where the product is taken over all primes p.

This follows from Theorem 1 by a formal computation. (For the detailed proofs of Theorems 1 and 2, see [8].)

Suppose that the above f is a cusp form, and define A_n by

$$(2.1)\qquad \sum_{n=1}^{\infty} A_n n^{-s} = \Pi_p[1 - \omega_p p^{-s} + \chi(p)^2 p^{k-2-2s}]^{-1}.$$

Obviously this is independent of t. Then we consider the Mellin (inverse) transform of this Dirichlet series:

$$(2.2)\qquad F(z) = \sum_{n=1}^{\infty} A_n e^{2\pi inz} \qquad (z \in H).$$

Now this function F is an ordinary modular form of weight k - 1. We state this fact in a more general form as

MAIN THEOREM: *Let* $f(z) = \sum_{n=1}^{\infty} a(n)e^{2\pi inz}$ *be an element of* $S_k(N, \chi)$ *with* $k > 3$, *and* t *a square-free positive integer. Define a character* χ_t *modulo* tN *by*

$$\chi_t(m) = \chi(m)(\tfrac{-1}{m})^{\lambda}\ (\tfrac{t}{m})$$

with $\lambda = (k - 1)/2$, *and a function* F_t *on* H *by*

$$F_t(z) = \sum_{n=1}^{\infty} A_t(n)e^{2\pi inz},$$

$$\sum_{n=1}^{\infty} A_t(n)n^{-s} = (\sum_{m=1}^{\infty} \chi_t(m)m^{\lambda-1-s})\ (\sum_{m=1}^{\infty} a(tm^2)m^{-s}).$$

Suppose that f *is a common eigen-function of* $T(p^2)$ *for all prime factors* p *of* N *not dividing the conductor of* χ_t. *Then* F_t *is an ordinary modular form of weight* k - 1 *satisfying*

$$F_t((az + b)/(cz + d)) = \chi(d)^2(cz + d)^{k-1}F_t(z)$$

for all $\begin{bmatrix} a & b \\ c & d \end{bmatrix} \in \Gamma_0(N_t)$, *with a positive integer* N_t. *Moreover* F_t *is a cusp form if* $k > 5$.

The integer N_t depends on N, χ, t, and f; it can be determined easily, but in a somewhat complicated manner. Here we note only that:

(i) all prime factors of N_t divide N;

(ii) $N_t = N/2$ if t is odd and every prime factor of N divides the conductor of χ_t.

COROLLARY: *The notation being as in Theorem* 2, *suppose that* $k > 3$ *and* f *is a cusp form. Define a function* F *on* H *by* (2.1) *and* (2.2). *Then* F *is an ordinary modular form of weight* $k - 1$ *satisfying*

$$F((az + b)/(cz + d)) = \chi(d)^2(cz + d)^{k-1}F(z) \quad \text{for all } \begin{bmatrix} a & b \\ c & d \end{bmatrix} \in \Gamma_0(N_0),$$

where N_0 *is the greatest common divisor of the integers* N_t *for all square-free* t *such that* $a_t \neq 0$. *If* $k > 5$, F *is a cusp form.*

EXAMPLE: $S_9(\Delta_0(4))$ is one-dimensional, and spanned by $f(z) = \theta(z)^{-3}\eta(2z)^{12}$. Then the corresponding F is a cusp form of weight 8 with respect to $\Gamma_0(2)$. Actually $F(z) = (\eta(z)\eta(2z))^8$. If $f(z) = \sum_{n=1}^{\infty} a(n)e^{2\pi inz}$ and $F(z) = \sum_{n=1}^{\infty} A(n)e^{2\pi inz}$, then

$$A(2) = a(4),$$

$$A(p) = a(p^2) + p^3 \qquad (p : \text{odd prime}).$$

Jacquet and Langlands [2] established a theory which connects Hecke theory with representation theory of GL(2) over local fields and the adeles. Now we are led to a natural question: "Can one construct a similar theory which connects the above results with representations of coverings of GL(2)?" For the discussion of some other open questions, the reader is referred to [8, §4].

3. Sketch of the proof of the main theorem

The idea of the proof is as follows. We first prove the analytic continuation and the functional equation of

$$(3.1)\qquad \sum_{n=1}^{\infty} \Psi(n)A_t(n)n^{-s}$$

with any primitive character Ψ whose conductor is either 1 or a prime not dividing tN. Then the criterion of Weil [13] guarantees that the function F_t is a modular form of the desired type. Now to prove the functional equation of (3.1), we adapt the method of Rankin [7] to the present situation.

Let us now explain this in more detail in the special case $t = 1$. Let r be either 1 or a prime not dividing N, and Ψ a primitive character modulo r. Define χ_1 and $A_1(n)$ as in the main theorem with $t = 1$. Then

$$(3.2)\quad \sum_{n=1}^{\infty} \Psi(n)A_1(n)n^{-s} = \left(\sum_{m=1}^{\infty} \Psi(m)\chi_1(m)m^{\lambda-1-s}\right)\left(\sum_{m=1}^{\infty} \Psi(m)a(m^2)m^{-s}\right).$$

Let h_Ψ be as in (1.8). Looking at the Fourier expansions of f and h_Ψ, we see easily that

$$\int_0^1 f\,\overline{h}_\Psi dx = \sum_{m=1}^{\infty} \Psi(m)a(m^2)m^{\nu}e^{-4\pi m^2 y} \qquad (z = x + iy),$$

hence, at least formally,

$$(3.3)\quad \int_0^{\infty} \left\{\int_0^1 f\,\overline{h}_\Psi dx\right\} y^{s-1}dy = (4\pi)^{-s}\Gamma(s) \sum_{m=1}^{\infty} \Psi(m)a(m^2)m^{\nu-2s}.$$

The left hand side is the integral of $f\,\overline{h}_\Psi y^{s+1}\, y^{-2}dxdy$ on the domain

$$Y = \{z = x + iy \mid 0 \leq x < 1,\ 0 < y\},$$

which is a fundamental domain for H modulo the parabolic subgroup

$$\Gamma_\infty = \{\pm\left[\begin{smallmatrix}1 & m\\ 0 & 1\end{smallmatrix}\right] \mid m \in \mathbb{Z}\}$$

of $\Gamma_0(r^2N)$. Now put $B(z, s) = f\,\bar{h}_\Psi y^{s+1}$, and observe that

$$B(\gamma(z), s) = J(\gamma, z, s)B(z, s) \qquad \text{for all} \quad \gamma \in \Gamma_0(r^2N),$$

where

$$J(\gamma, z, s) = \Psi(d)\chi_1(d)(cz + d)^{\lambda-\nu}|cz + d|^{2\nu-1-2s} \qquad \text{for} \quad \gamma = \left[\begin{smallmatrix}a & b\\ c & d\end{smallmatrix}\right],$$

since $f \in S_k(N, \chi)$ and $h_\Psi \in M_{2\nu+1}(4r^2, \Psi_1)$. Therefore, if R is a set of representatives for $\Gamma_\infty\backslash\Gamma_0(r^2N)$, the above integral can be reduced to the form

$$(3.4)\quad \int_Y B(z, s)y^{-2}dxdy = \int_\Phi B(z, s)\ \{\textstyle\sum_{\gamma \in R} J(\gamma, z, s)\}\ y^{-2}dxdy,$$

where Φ is a fundamental domain for H modulo $\Gamma_0(r^2N)$.

Let W be the set of all couples $\{c, d\}$ of relatively prime integers such that $c \equiv 0 \pmod{r^2N}$, $c > 0$ if $c \neq 0$, $d = 1$ if $c = 0$. Then R can be chosen so that

$$R \ni \left[\begin{smallmatrix}a & b\\ c & d\end{smallmatrix}\right] \mapsto \{c, d\} \in W$$

is one-to-one. Therefore, putting

$$L(s) = \textstyle\sum_{n=1}^{\infty} \Psi(n)\chi_1(n)n^{-s},$$

we obtain

$$2\,L(2s-\lambda-\nu+1)\sum_{\gamma\in R} J(\gamma, z, s)$$

$$= \sum{}' \,\Psi(d)\chi_1(d)(cz+d)^{\lambda-\nu}|cz+d|^{2\nu-1-2s},$$

where $\sum'$ is extended over all $(c, d) \in r^2N\mathbf{Z} \times \mathbf{Z}$ excluding $(0, 0)$. Denote the last sum by $E(z, s)$. Then

$$2\,(4\pi)^{-s}\Gamma(s)\sum_{n=1}^{\infty} \Psi(n)A_1(n)n^{\nu-2s}$$

$$(3.5) \qquad = 2\,(4\pi)^{-s}\Gamma(s)L(2s-\lambda-\nu+1)\sum_{m=1}^{\infty} \Psi(m)a(m^2)m^{\nu-2s}$$

$$= \int_{\Phi} B(z, s)E(z, s)y^{-2}dxdy.$$

Now $E(z, s)$ is an "Eisenstein-Epstein" series, which can be continued analytically to the whole s-plane, and satisfies a functional equation under $s \mapsto \lambda + \nu - s$ (of course with a suitable Γ-factor). Also we can study the behavior of $E(z, s)$ at each cusp, and show that the last integral is convergent for all s. Thus the above "formal" computation is actually valid, and (3.5), multiplied by a suitable Γ-factor, is an entire function in s. Then, as we expect, the transformation $s \mapsto \lambda + \nu - s$, combined with $z \mapsto -1/r^2Nz$, yields the desired functional equation for (3.1). But this is not so straightforward as it looks. Actually $E(z, \lambda + \nu - s)$ is not so simple as the original $E(z, s)$. To overcome this difficulty, we need a rather lengthy computation and a somewhat tricky reduction process, of which the details are given in [8].

4. Eisenstein series and an application

The main theorem controls the Fourier coefficients

$$\{a(tm^2) \mid m = 1, 2, \dots\}$$

Shi-34

for each fixed square-free t, but does not give any information about the mutual relationship of the numbers a(t) for all square-free t. Is there any principle by which these a(t) are governed? This question seems very difficult, but one can at least treat the same question for Eisenstein series. For simplicity let us consider here only the case of level 4, in which we have exactly two Eisenstein series for each weight k/2, if $k \geq 5$. One of them has the following form:

$$E_k(z) = \sum \left(\frac{-m}{n}\right) \varepsilon_n^k (nz + m)^{-k/2},$$

where ε_n is as in (1.3), and the sum is extended over all couples {m, n} of integers such that $(m, n) = 1$, $n > 0$, $n \equiv 1 \pmod 2$. Then $M_k(\Delta_0(4))$ is spanned, modulo $S_k(\Delta_0(4))$, by $E_k(z)$ and $E_k(-1/4z)\, z^{-k/2}$.

To obtain the Fourier coefficients of E_k, we use a well-known formula

$$\sum_{h=-\infty}^{\infty} (z + h)^{-a} = (2\pi)^a e^{-\pi i a/2} \Gamma(a)^{-1} \sum_{n=1}^{\infty} n^{a-1} e^{2\pi i n z}$$

($a > 0$, $z \in H$), and obtain

$$E_k(z) = (2\pi)^{k/2} e^{-\pi i k/4} \Gamma(k/2)^{-1} \sum_{n=1}^{\infty} a(n) e^{2\pi i n z},$$

where

$$a(n) = n^{k/2-1} \sum_{m>0,\ \text{odd}} \varepsilon_m^k\, m^{-k/2} \left(\sum_{0<r<m} \left(\frac{-r}{m}\right) e^{2\pi i n r/m} \right).$$

After some transformations, we can show that, for every square-free t,

$$\sum_{n=1}^{\infty} a(tn^2) n^{-s} = \frac{a(t)}{1 - 2^{k-2-s}} \prod_{p\,:\,\text{odd}} \frac{1 - \left(\frac{t}{p}\right)\left(\frac{-1}{p}\right)^{\lambda} p^{\lambda-1-s}}{(1 - p^{-s})(1 - p^{k-2-s})},$$

where $\lambda = (k - 1)/2$. Put

$$\phi_t(m) = \left(\frac{t}{m}\right)\left(\frac{-1}{m}\right)^{\lambda},$$

$$L(s, \phi_t) = \sum_{m=1}^{\infty} \phi_t(m)m^{-s}.$$

Then $$a(t) = t^{k/2-1}(1 - 2^{1-k})^{-1}\zeta(k - 1)^{-1}L((k - 1)/2, \phi_t),$$

where ζ is the Riemann zeta function. (The occurrence of $L((k-1)/2, \phi_t)$ is already noticed in Petersson [4] and Maass [3]. One may also refer to the results of Eisenstein on the number of representations of an integer by five or seven squares, and Hardy's approach to this topic.)

Finally we mention an application of the method of §3 to a certain type of Dirichlet series. Let

$$g(z) = \sum_{n=1}^{\infty} c(n)e^{2\pi inz}$$

be a cusp form of weight w and level 1 (so that $g((az + b)/(cz + d)) = (cz + d)^w g(z)$ for all $\begin{bmatrix} a & b \\ c & d \end{bmatrix} \in SL_2(\mathbf{Z})$, and w is even, $w \geq 12$). Suppose that $c(1) = 1$ and g is a common eigen-function of all Hecke operators. Then we have an Euler product

$$\sum_{n=1}^{\infty} c(n)n^{-s} = \prod_p (1 - c(p)p^{-s} + p^{w-1-2s})^{-1}.$$

Now decompose each Euler factor in the form

$$1 - c(p)p^{-s} + p^{w-1-2s} = (1 - u_p p^{-s})(1 - v_p p^{-s}),$$

and define a new Euler product D(s) by

$$D(s) = \prod_p (1 - u_p^2 p^{-s})^{-1}(1 - v_p^2 p^{-s})^{-1}(1 - u_p v_p p^{-s})^{-1}.$$

Note that $u_p v_p = p^{w-1}$. We can easily verify that

$$\zeta(2s - 2w + 2)\sum_{n=1}^{\infty} c(n)^2 n^{-s} = \zeta(s - w + 1)D(s),$$

$$\zeta(2s - 2w + 2)\sum_{n=1}^{\infty} c(n^2) n^{-s} = D(s).$$

Notice the distinction between $c(n)^2$ and $c(n^2)$. Rankin [7] showed that the function

$$R^{*}(s) = (2\pi)^{-2s}\Gamma(s)\Gamma(s - w + 1)\zeta(s - w + 1)D(s)$$

can be continued to a meromorphic function on the whole s-plane, which is holomorphic except for simple poles at $s = w$ and $w - 1$; further it satisfies

(4.1) $$R^{*}(2w - 1 - s) = R^{*}(s).$$

Now we consider, instead of R^{*}, the following function

$$R(s) = 2^{-s}\pi^{-3s/2}\Gamma(s)\Gamma\big((s - w + 2)/2\big)D(s).$$

Obviously this is meromorphic on the whole plane, and if we combine (4.1) with the functional equation of $\zeta(s - w + 1)$, we obtain

$$R(2w - 1 - s) = R(s).$$

THEOREM 3: $R(s)$ <u>is holomorphic on the whole plane</u>.

To prove this, we observe that

$$\int_Y g\,\overline{\theta}\, y^{s-1}dxdy = 2\,(4\pi)^{-s}\Gamma(s)\sum_{m=1}^{\infty} c(m^2)m^{-2s},$$

which is similar to (3.3). Then, by the same principle as in (3.4), we transform it to an integral on a fundamental domain for $H/\Gamma_0(4)$. This time we obtain, instead of $E(z, s)$, a series of the form

$$\sum_{\gamma \in R} j(\gamma, z)^{2w-1}|cz + d|^{-2s-2} \qquad (\gamma = \begin{bmatrix} a & b \\ c & d \end{bmatrix})$$

with a set R of representatives for $\Gamma_\infty \backslash \Gamma_0(4)$. This belongs to the series of the type considered by Siegel in [10], [11]. Therefore, checking carefully the convergence of the integral, we obtain the above theorem. The detail of the proof will appear in [9].

REFERENCES

[1] E. HECKE: Herleitung des Euler-Produktes der Zetafunktion und einiger L-Reihen aus ihrer Funktionalgleichung, Math. Ann. 119 (1944), 266-287 (= Werke, 919-940).

[2] H. JACQUET and R. P. LANGLANDS: Automorphic forms on GL(2), Lecture notes in mathematics, 114, Springer, 1970.

[3] H. MAASS: Konstruktion ganzer Modulformen halbzahliger Dimension mit ϑ-Multiplikatoren in einer und zwei Variablen, Hamburg Abh. 12 (1937), 133-162.

[4] H. PETERSSON: Über die Entwicklungskoeffizienten der ganzen Modulformen und ihre Bedeutung für die Zahlentheorie, Hamburg Abh. 8 (1931), 215-242.

[5] H. PETERSSON: Über die systematische Bedeutung der Eisensteinschen Reihen, Hamburg Abh. 16 (1949), 104-130.

[6] H. PETERSSON: Über Betragmittelwerte und die Fourier-Koeffizienten der ganzen automorphen Formen, Archiv der Math. 9 (1958), 176-182.

[7] R. A. RANKIN: Contributions to the theory of Ramanujan's function $\tau(n)$ and similar arithmetical functions I, II, Proc. Cambridge Phil. Soc. 35 (1936), 351-372.

[8] G. SHIMURA: On modular forms of half integral weight, to appear in the Ann. of Math.

[9] G. SHIMURA: On the holomorphy of certain Dirichlet series, in preparation.

[10] C. L. SIEGEL: Die Funktionalgleichungen einiger Dirichletscher Reihen, Math. Z. 63 (1956), 363-373 (= Abh. III, 228-238).

[11] C. L. SIEGEL: A generalization of the Epstein zeta function, J. Ind. Math. Soc., 20 (1956), 1-10 (= Abh. III, 239-248).

[12] A. WEIL: Sur certains groupes d'opérateurs unitaires, Acta Math. 111 (1964), 143-211.

[13] A. WEIL: Über die Bestimmung Dirichletscher Reihen durch Funktionalgleichungen, Math. Ann. 168 (1967), 149-156.

[14] K. WOHLFAHRT: Über Operatoren Heckescher Art bei Modulformen reeller Dimension, Math. Nachr. 16 (1957), 233-256.

THE BASIS PROBLEM FOR MODULAR FORMS
AND THE TRACES OF THE HECKE OPERATORS

BY M. EICHLER

International Summer School on Modular Functions

Antwerp 1972

CONTENTS

INTRODUCTION

In the following article we consider holomorphic modular forms with respect to the congruence subgroup $\Gamma_0(N) = \{(\begin{smallmatrix} a & b \\ c & d \end{smallmatrix}) \in \Gamma : c \equiv 0 \bmod N\}$ of the modular group Γ. Of course the holomorphy condition includes the cusps. By $S_k(\Gamma_0(N),\chi)$ we denote the space of these forms of weight k and character χ, i.e. those satisfying

$$f(\frac{az+b}{cz+d})\ (cz+d)^{-k} = \chi(a)\ f(z)$$

for all substitutions of $\Gamma_0(N)$, which vanish in the cusps.

Our main problem, the basis problem, is to give bases of linearly independent forms of these spaces which are arithmetically distinguished and whose Fourier series are known or easy to obtain.

The solution rests on an arithmetic counterpart to Hecke's theory in the arithmetic of quaternion algebras. It has been initiated by Brandt, and we may reasonably call the analogues of Hecke's T(n) the Brandt matrices. The link between them and the T(n) is the fact that both generate isomorphic semisimple rings with the same traces. The determination of the traces is therefore our chief concern.

In I we give a proof of the fact that theta series (generalized by Schoeneberg) are modular forms. It differs from the classical proof by Hermite in the respect that the (finite) modular congruence group $\Gamma/\Gamma(N)$ is used in an essential way. We think that this procedure seems less artificial. In this connection we meet the problem of the representations of this group which is today only partly solved.

In II we develop the arithmetic of quaternion algebras, concluding with the computation of the traces of the Brandt matrices. In III the traces of the Hecke operators in the space of modular forms with respect to the congruence subgroups $\Gamma_0(N)$, with a square-free N, are determined. This requires at first some function theoretic considerations on an analogue of Greens function and secondly the counting of the fixed points of the modular correspondences. In this second part we use essential results of II. In IV the previous results are combined to solve the basis problem for square-free N. The proof of the final theorem in IV seems unusually long. Perhaps one could state that that it amounts to establishing a certain connection between definite quaternion algebras and the matrix algebra. Accepting this view, one would further be led to the feeling that there may exist another standpoint from which the theory developed so far would look much easier. Therefore our final theorem is as much a challenge to further work on the problem as an information on existing facts.

We have to excuse ourselves for considering only square-free N, although Hijikata (quotation [4] in III) has already treated the case of a general N. We hoped to ease the work both for the reader and for the author, even if the gain is not very large.

The congruence subgroups $\Gamma_0(N)$ of the modular group are most important because of their applications to theta functions and zeta functions. Taken on their own merit, the principal congruence subgroups

$$\Gamma(N) = \{G \in \Gamma : G \equiv \pm\begin{pmatrix}1 & 0\\ 0 & 1\end{pmatrix} \bmod N\}$$

may be even more interesting, particularly because of their connection with the factor groups

$$M(N) = \Gamma/\Gamma(N).$$

The determination of the traces of the T(n) for such modular forms may even be easier than for $\Gamma_0(N)$, as soon as the representations of $M(N)$ are known. In the appendix we note some facts on these forms, incomplete as they are today.

What concerns the presentation we try to compromise between the style of a text book and a research paper. The arithmetic of quaternion algebras may not be familiar to a modern reader. But if he is only experienced in algebraic number theory, a brief summary of the basic facts may enable him to interpolate the rest. A similar attitude has been adopted with respect to the function-theoretic background of the determination of the traces of Hecke's T(n).

We are pleased to express our thanks to H.Hijikata who helped to check our results against his recent work which is due to appear soon.

Eich-6

Chapter I: THE THETA SERIES

§1. Definitions

Let F be the matrix of a positive definite quadratic form in an even number 2k of variables and

$$F = S^t S \tag{1}$$

with a real matrix S. The coefficients of F are assumed to be rational integers; those in the diagonal as even integers. The least natural integer N for which $N F^{-1}$ has the same properties is called the level (Stufe) of F.

We will use all solutions of

$$F r \equiv 0 \bmod N \tag{2}$$

in rational integral column vectors r. The number of solutions which are incongruent mod N is equal to the determinant $|F|$.

Furthermore, we let $p_l(x) = p_l(x_1, \ldots, x_{2k})$ be a homogeneous polynomial of degree l satisfying the Laplace differential equation

$$\Delta\, p_l(x) = 0.$$

It is known that every polynomial solution of this equation is uniquely expandable in a series

$$f(x) = \sum p_l(x), \tag{3}$$

and that for $l_1 \neq l_2$ the $p_l(x)$ are orthogonal, namely

$$\int p_{l_1}(x)\, p_{l_2}(x) d\omega = 0, \tag{4}$$

integrated over the surface of a sphere $\sum x_i^2 = \rho^2$.

To a triple consisting of a quadratic form

$$F[x] = x^t F\, x,$$

a solution r of (2), and a polynomial $p_l(x)$ we associate the (generalized) theta series

(5) $\theta(z,r) = (-1)^{l}\theta(z,-r) = \sum_{n \in \mathbb{Z}^{2k}} p_l(S(n + N^{-1}r))\, e^{\pi iz F[n+N^{-1}r]}$.

It is the main task of I to establish the $\theta(z,r)$ as modular forms of weight $k+l$, of level N, and of a certain character χ.

Here and later we treat the elements $\begin{pmatrix} a & b \\ c & d \end{pmatrix}$ of the full modular group Γ as linear operators on functions defined by

(6) $$\phi(z) \begin{bmatrix} a & b \\ c & d \end{bmatrix}^{-k} = \phi\left(\frac{az+b}{cz+d}\right)(cz+d)^{-k}.$$

§2 The functional equations

The following functional equation is an immediate consequence of (2):

(7) $$\theta(z,r) \begin{bmatrix} 1 & 1 \\ 0 & 1 \end{bmatrix}^{-k-l} = \theta(z+1,r) = e^{\pi i N^{-2} r^t F r}\theta(z,r).$$

The next one is

(8) $$\theta(z,r) = \frac{i^k}{\sqrt{|F|}} z^{-k-l} \sum_{n \in \mathbb{Z}^{2k}} p_l(S^{-t}(m))\, e^{-\pi i z^{-1} F^{-1}[m] + 2\pi i N^{-1} r^t m}$$

(where S^{-t} means $(S^t)^{-1}$).

Proof: The function (5): $\phi(x)$ where $N^{-1}r$ is replaced by $N^{-1}r + x$ with a variable column vector x is holomorphic and periodic with periods 1 in all components x_i of x and therefore expandable in a Fourier series

(9) $$\phi(x) = \sum c_m\, e^{2\pi i m^t x}.$$

The coefficients are

$$c_m = \int_0^1\!\!\cdots\!\int \phi(x)\, e^{-2\pi i m^t x}\, dx_1 \dots dx_{2k}.$$

Because of the uniform convergence of the series the integration and the summation can be interchanged, which leads to

$$c_m = \sum_n \int_0^1\!\!\cdots\!\int p_l(S(n+x+N^{-1}r))\, e^{\pi i z F[n+x+N^{-1}r] - 2\pi i m^t(n+x)}.$$

Eich-8

We may write $n + x = x'$ and combine summation and integration to

$$c_m = \int_{-\infty}^{+\infty}\!\!\cdots\!\int p_1(S(x'+N^{-1}r))\, e^{\pi iz\, F[x'+N^{-1}r] - 2\pi i m^t x'} dx_1' \ldots dx_{2k}'.$$

In the next step we write

$$x' - z^{-1}F^{-1}m + N^{-1}r = x''$$

and integrate over all x_1'' from $-\infty$ to $+\infty$ which is allowable because of the Cauchy theorem and the vanishing of the integrand in the infinite. In the exponent we have

$$\pi izF\,[x'+N^{-1}r] - 2\pi i m^t x' =$$

$$= \pi izF\,[x'-z^{-1}F^{-1}m + N^{-1}r] - \pi iz^{-1}F^{-1}[m] + 2\pi iN^{-1}m^t r,$$

and c_m becomes

$$c_m = e^{-\pi iz^{-1}F^{-1}[m] + 2\pi iN^{-1}m^t r} \int_{-\infty}^{+\infty}\!\!\cdots\!\int p_1(S(x''+z^{-1}F^{-1}m))$$

$$e^{\pi izF[x'']} dx_1'' \ldots dx_{2k}''.$$

In order to evaluate the integral we put $z = it$ with a positive real t and introduce new variables $y = \sqrt{t}Sx''$. Then

$$p_1(S(x''+z^{-1}F^{-1}m)) = t^{-1}p_1(\sqrt{t}\; y - iS^{-t}m).$$

We expand the function $p_1(\sqrt{t}\; y-iS^{-t}m)$ in the manner (3), the term $p_0(y)$ being $p_1(-iS^{-t}m) = (-i)^l p_1(S^{-t}m)$. Introducing polar coordinates in the integral and observing (4) we obtain

$$c_m = \frac{(-i)^l}{\sqrt{|F|}} t^{-k-l} p_1(S^{-t}m)\; e^{-\pi iz^{-1}F^{-1}[m] + 2\pi iN^{-1}m^t r}\; \omega_{2k} \int_0^\infty e^{-\pi\rho^2} \rho^{2k-1}\, d\rho$$

where ω_{2k} is the surface of the 2k-dimensional unit sphere, namely

$$\omega_{2k} = \frac{2\pi^k}{(k-1)!}, \text{ and } \int_0^\infty e^{-\pi\rho^2}\rho^{2k-1}\, d\rho = \frac{(k-1)!}{2\pi^k}.$$

At last we reintroduce z and get

$$c_m = \frac{(-i)^k}{\sqrt{|F|}} z^{-k-1} P_1(S^{-t}m)\, e^{-\pi i z^{-1}F^{-1}[m] + 2\pi i N^{-1} r^t m}.$$

Using (9) with these Fourier coefficients and inserting $x = 0$ we obtain (8).

The right hand side of (8) can be transformed by putting

$$m = Fn + s$$

where n runs over $\mathbb{Z}^{2k}$ and s over a system of representatives of the factor group $\{m\} / \{Fn\}$. To each such s we can associate the solution

$$r' = N F^{-1} s$$

of (2), and vice versa. Therefore we can also write $m = F(n + N^{-1}r')$, and (8) becomes (with $-z^{-1}$ instead of z)

$$(10)\quad \theta(z,r)\left[\begin{pmatrix} 0 & 1 \\ -1 & 0\end{pmatrix}\right]^{-k-1} = \theta(-z^{-1},r)(-z)^{-k-1} = \frac{(-i)^k}{\sqrt{|F|}} \sum e^{2\pi i N^{-2} r^t F r'}\theta(z,r'),$$

to be summed over a complete system r' of solutions of (2) mod N.

§3 Representations of the modular congruence group

(7) and (10) describe the behaviour of the $\theta(z,r)$ under the operators (translation and inversion)

$$T = \begin{pmatrix} 1 & 1 \\ 0 & 1\end{pmatrix}, \qquad J = \begin{pmatrix} 0 & 1 \\ -1 & 0\end{pmatrix}$$

which generate the whole modular group. Next we ask for the behaviour under an element

$$R(d) \equiv \begin{pmatrix} d^{-1} & 0 \\ 0 & d\end{pmatrix} \bmod N$$

of Γ. Such an element is

$$(11)\qquad R(d) = T^{d^{-1}} J^{-1} T^{d} J^{-1} T^{d^{-1}} J^{-1}$$

where d^{-1} means an integer satisfying $d\, d^{-1} \equiv 1 \bmod N$. Apparently $R(d)$, as an operator on the $\theta(z,r)$, does not change when d, d^{-1} are replaced by other integers in the same residue classes mod N. A

Eich-10

straightforward computation shows

$$\theta(z,r)\,[R(d)]^{-k-1} =$$

$$= \frac{i^k}{|F|^{3/2}} \sum_{r_1,r_2,r'} e^{\pi i N^{-2}(d^{-1}r^tFr+2r^tFr_1+dr_1^tFr_1+2r_1^tFr_2+d^{-1}r_2^tFr_2+2r_2^tFr')} \cdot$$

$$\cdot\theta(z,r'),$$

where r_1, r_2, r' run each over a system of solutions of (2). The exponent is N^{-2} times

$$\pi i d\, F[d^{-1}r + r_1 + d^{-1}r_2] + 2\pi i(r'- d^{-1}r)^tF\, r_2.$$

Instead of r_1, r_2, r' we let r_2, r', $d^{-1}r + r_1 + d^{-1}r_2$ run over the solutions of (2). The summation over r_2 is carried out at first and yields always 0, except when $r' - d^{-1}r \equiv 0 \bmod N$, and then it yields $|F|$ because (2) has so many solutions. Therefore

$$(12)\qquad \theta(z,r)\,[R(d)]^{-k-1} = \chi(d)\theta(z,d^{-1}r)$$

with

$$(13)\qquad \chi(d) = \frac{i^k}{\sqrt{|F|}} \sum e^{\pi i N^{-2}d\, F[r]}.$$

<u>Proposition 1.</u> *For* $\varepsilon = \pm 1$ *and every matrix of the form*

$$\begin{pmatrix} a & b\\ c & d\end{pmatrix} \equiv \varepsilon\begin{pmatrix} 1 & 0\\ 0 & 1\end{pmatrix} \bmod N, \qquad ad-bc = 1$$

the functional equations hold:

$$(14)\qquad \theta(z,r)\begin{bmatrix} a & b\\ c & d\end{bmatrix}^{-k-1} = \varepsilon^{k+1}\theta(z,r).$$

In other words, the theta series $\theta(z,r)$ *are modular forms with respect to the prinicipal congruence subgroup* $\Gamma(N)$ *mod* N.

<u>Proof</u>: Equation (14) with $\begin{pmatrix} a & b\\ c & d\end{pmatrix} = \varepsilon\begin{pmatrix} 1 & 0\\ 0 & 1\end{pmatrix}$ follows immediately from the definition (6).

(12) exhibits the $R(d)$ as operators which form a group isomorphic to $(\mathbb{Z}/N\mathbb{Z})^*$, and $\chi(d)$ is a character of it. T generates a group of

operators isomorphic to $(\mathbb{Z}/N\mathbb{Z})^*$.

The following equation is easily verified by means of (7) and (12):

$$R(d)^{-1}\, T\, R(d) = T^{d^2}, \tag{15}$$

and from (10), (12), (13) follows ($\chi(d) = \chi(d^{-1})$ and)

$$J^{-1}R(d)\, J = R(d^{-1}) \tag{16}$$

while

$$J^2 = \begin{pmatrix} -1 & 0 \\ 0 & -1 \end{pmatrix} \tag{17}$$

is a consequence of (10), (12), (14) and the fact that $p_1(x)$ has the degree 1. In (15) - (17), the elements have to be understood as operators on the $\theta(z,r)$.

What has to be proved is that T, J as operators generate a group which is isomorphic to the modular congruence group SL(2, $\mathbb{Z}/N\mathbb{Z}$). Let $N = \Pi p^\lambda$ be the decomposition into different primes. Then the operators J, $T^{N\, p^{-\lambda}}$, and $R(d_p)$ with $d_p \equiv 1 \bmod N\, p^{-\lambda}$ satisfy the same relations between each other with the only difference that the groups generated by $R(d_p)$ and $T^{N\, p^{-\lambda}}$ are the analogue groups with p^λ instead of N. Therefore it suffices to treat the case $N = p^\lambda$.

This we begin with showing that any operator $W = \begin{pmatrix} a & b \\ c & d \end{pmatrix}$ with $ad - bc = 1$ can be written in the form

$$W = J^e\, T^a\, J\, T^b\, J^{e'}\, R(c), \qquad e \text{ and } e' = 0 \text{ or } 1. \tag{18}$$

It is known that always

$$W = T^{a_1}\, J\, T^{a_2} \dots T^{a_r}\, J$$

which shows that (18) is possible for r = 1 or 2. Let (18) hold for r-1, then

$$W = T^a\, J\, T^b\, J\, T^c\, J^{e'}\, R(d).$$

Eich-12

If $b \not\equiv 0 \bmod p$ we can transform this using (11) into

$$W = R\, T^{a-b^{-1}}\, R(b)\, J\, T^{c-b^{-1}}\, J^{e'}\, R(d)$$

and further into (18) by means of (15) and also (16). If $b \equiv 0 \bmod p$, we apply (11) two times with $d = 1$ and $d = b-1$ (and (17)):

$$W = T^{a-1}\, J\, T^{-1-(b-1)^{-1}}\, R(b-1)\, J\, T^{c-(b-1)^{-1}}\, J^{e'}\, R(d).$$

The factor $R(b-1)$ can be combined with $R(d)$ by virtue of (15) and (16), which leads to

$$W = T^{a-1}\, J\, T^{b'}\, J\, T^{c'}\, J^{e'}\, R(d')$$

where again $b' \equiv 0 \bmod p$. We repeat this procedure a times and arrive at one of the forms (18).

Now we assume $W \equiv \begin{pmatrix} 1 & 0 \\ 0 & 1 \end{pmatrix} \bmod N$. The same congruence holds for $J^{-1}W\, J$. Therefore we only need discuss the case with $e = 0$. Then $e' = 1$ and

$$W = \begin{pmatrix} (ab-1)c^{-1} & -ac \\ bc^{-1} & -c \end{pmatrix} \equiv \begin{pmatrix} 1 & 0 \\ 0 & 1 \end{pmatrix} \bmod N.$$

This implies $a \equiv b \equiv 0 \bmod p^{\lambda}$ and $c \equiv -1 \bmod p^{\lambda}$. But T^a and T^b are then the unit operator, and because of (17) also $J^2R(c)$. This completes the proof.

As a consequence of (12) and proposition 1 we have

(19) $$\theta(z,r) \left[\begin{matrix} a & b \\ c & d \end{matrix}\right]^{-k-1} = \chi(d)\theta(z,d^{-1}r),\quad \begin{pmatrix} a & b \\ c & d \end{pmatrix} \equiv \begin{pmatrix} d^{-1} & 0 \\ 0 & d \end{pmatrix} \bmod N,$$

and this must include (14) or

(20) $$\chi(-1) = (-1)^{k+1}.$$

Of course, we are interested in the character defined by (13).

Proposition 2. *$\chi(d)$ is a character of the group $(\mathbb{Z}/N\mathbb{Z})^*$. It is $\chi(d)=1$ if d is congruent mod N to the norm of a similarity transformation of F. For instance, this is the fact for d = p, a prime, with the Legendre symbol*

$$\left(\frac{(-1)^k |F|}{p}\right) = 1.$$

Thus $\chi(d)$ can only assume the values ±1.

Quite generally we have

$$\chi(d) = (\mathit{sign}\ d)^{k+1} \left(\frac{(-1)^k |F|}{|d|}\right). \tag{21}$$

These facts are almost obvious apart from (21). By (13), $\chi(d)$ is a Gauss sum which can be determined in the well known way (see for instance [1].

If $(-1)^k |F|$ is a square, $\chi(d)$ is always 1. So far we have defined $\chi(d)$ only for $(d, N) = 1$. At later occasions we shall put $\chi(d) = 0$ if $(d,N) > 1$.

§4 Concluding remarks

That the $\theta(z,r)$ are modular forms has been discovered by Hermite [3] if the $p_1(x) = 1$. The case for arbitrary $p_1(x)$ was first treated by Schoeneberg [5]. Kloosterman [4] made an extensive study of more general theta series (but with $p_1(x) = 1$) with the object of determining the representations of the modular congruence group $SL(2, \mathbb{Z}/N\mathbb{Z})$. His results are complicated and cannot be described briefly. Three questions remain open: 1) whether all representations of these groups are thus obtained, 2) whether such generality is actually needed for the purpose, 3) whether there are identically satisfied linear relations between the theta series used. We think it necessary and possible, though not quite easy to simplify and to reorder Kloosterman's work.

As we have seen in the proof of proposition 1, (7) and (10) define

Eich-14

a representation of the modular congruence group SL(2, $\mathbb{Z}/N\mathbb{Z}$). It is not known what representations occur in this way. Here are a few examples.

1) $|F| = N = q$, an odd prime and $l = q^{k-1}$ Evidently $\theta(z,r) = \theta(z,-r)$. Therefore we have $\frac{q+1}{2}$ different theta series for a particular F. The degree of the representation is $\frac{q+1}{2}$. According as k is even or odd, $q \equiv \pm 1 \bmod 4$. In the former case we have to use the tables of representations given by Frobenius and reproduced by Hecke [2], in the latter case the tables given by Schur [7]. There exist exactly 2 irreducible representations of this degree, and none of smaller degree except the identical representation. Therefore our representation is irreducible, and all formally different $\theta(z,r)$ are linearly independent.

2) The case $k = 2$, $|F| = q^2$, $N = q$, an odd prime, has been treated by Schoeneberg [6], though in a different setting. His representations split into $G^{(i)}_{q-1}$, $G_{\frac{q-1}{2}}$, $G'_{\frac{q-1}{2}}$ in Hecke's notation.

Shimura [8] renewed Hermite's method, applying it to theta series in an arbitrary number n of variables, and with polynomials $p_l(x)$ with $l = 1$.

Literature

[1] M. Eichler, Einführung in die Theorie der algebraischen Zahlen und Funktionen, Birkhäuser Verlag Basel 1963, p. 64-65.
English translation, Academic Press, New York and London 1966, p. 50-51.

[2] E. Hecke, Math. Werke, Göttingen 1959 (Vandenhoek and Ruprecht), p. 529.

[3] Ch. Hermite, Sur quelques formules relative à la thransformations des fonctions elliptiques, Journal de Liouville (e. série) 3 (1858), p. 26.

[4] H. D. Kloosterman, The behaviour of general theta functions under the modular group and the characters of binary modular congruence groups I, II, Annals of Maths. 47 (1946), p. 317-447.

[5] B. Schoeneberg, Das Verhalten von mehrfachen Thetareihen bei Modulsubstitutionen, Math. Annalen 116 (1939)

[6] B. Schoeneberg, Über die Quaternionen in der Theorie der elliptischen Modulfunktionen, J. reine angew. Math. 193 (1954), p. 84-93.

[7] I. Schur, Untersuchungen über die Darstellungen der endlichen Gruppen durch gebrochen lineare Substitutionen, J. reine angew. Math. 132 (1907), p. 85-137.

[8] G. Shimura, On modular forms of half integral weight, to appear.

Eich-16

Chapter II: QUATERNION ALGEBRAS

§1. Summary of the ideal theory

Let Φ be a quaternion algebra over Q. We begin with considering modules M in Φ, namely finite $\mathbf{Z}$-modules of rank 4. They are always free. For a prime number p we form

$$\mathbf{Z}_p = \{\tfrac{a}{b}: a, b \in \mathbf{Z} \text{ and } (b,p) = 1\}.$$

The $\mathbf{Z}_p$-modules $M_p = \mathbf{Z}_p M$ are called the local components of M. We have always

(1) $$M = \bigcap_p M_p$$

and, N being another module,

(2) $$M_p = N_p \quad \text{for almost all p.}$$

Let N be a module and M_p a collection of finite $\mathbf{Z}_p$-modules of rank 4 satisfying (2). Then the intersection (1) is a finite $\mathbf{Z}$-module of rank 4.

An order is a ring and a finite $\mathbf{Z}$-module. Because of the finiteness condition, an order is contained in one or more maximal orders. An order I is maximal if and only if all p-adic components I_p are maximal orders (with respect to $\mathbf{Z}_p$). If I is an order, almost all I_p are maximal.

Let $\overline{Q}_p$, $\overline{\mathbf{Z}}_p$ be the closures of Q, $\mathbf{Z}_p$ under the p-adic topology. Then $\overline{\Phi}_p = Q_p\Phi$ is isomorphic to the matrix algebra M(2, Q_p) for almost all p. By D we denote the product of all primes for which $\overline{\Phi}_p$ is not the matrix algebra. The number of such primes is even or odd according as Φ is an indefinite or definite algebra, in other words if $\overline{\Phi}_\infty \cong M(2,R)$ or not.

For the p dividing D, $\overline{\Phi}_p$ and Φ contain exactly one maximal order $\overline{I}_p$ and I_p respectively. We are only interested in I_p. There exists exactly one prime I_p-ideal P_p. It has the properties

$$I_p P_p = P_p I_p = P_p, \quad P_p^2 = I_p p.$$

We will consider orders of the following properties and call them <u>orders of invariant</u> H:

a) I_p is maximal if $p \mid D$.

b) A square-free natural number H, prime to D will be arbitrarily fixed. For all $p \mid H$, I_p is isomorphic to the order of matrices $\begin{pmatrix} a & b \\ pc & d \end{pmatrix}$ with a, b, c, d $\in \mathbb{Z}_p$.

c) For all other p, I_p is isomorphic to the full matrix ring $M(2,\mathbb{Z}_p)$.

Let I be such an order. An I-left or -right ideal is a module M with the properties

$$IM = M \quad \text{or} \quad MI = M$$

and

(3) $M_p = I_p M_p$ or $= M_p I_p$ with M_p and $M_p^{-1} \in \Phi$ and almost always M_p and $M_p^{-1} \in I_p$. In the case of maximal orders (H = 1), (3) is superfluous and can be proved. An I-left (right) ideal M has a right (left) order I', namely the largest order I' with the property $MI' = M$ (or $I'M = M$). It is again of invariant H.

If the right order of M_1 is equal to the left order of M_2, the product $M_1 M_2$ is formed, namely the set of all finite sums $\sum m_{12} M_1 M_2$ with $m_{12} \in \mathbb{Z}$, $M_1 \in M_1$, $M_2 \in M_2$. In this case the local components have similar properties and satisfy

$$M_{1,p} M_{2,p} = (M_1 M_2)_p. \tag{4}$$

With the products of ideals thus defined, they form a <u>gruppoid</u>.

The <u>norm</u> of an ideal M is the g.c.d. of the norms of all elements of M. It satisfies the product rule

Eich-18

$$n(M_1)\, n(M_2) = n(M_1 M_2). \tag{5}$$

Formally, products $M_1 M_2$ can also be formed without the right order of M_1 coinciding with the left order of M_2. But then the associative law of the multiplication and the product rule for the norm are violated. Such a product is actually used, however, in showing that, for two orders, I_1, I_2 of invariant H, an ideal M exists with these left and right orders, namely $M = I_1 I_2$.

An ideal M is called <u>integral</u>, if it is contained in its left order. It is then also contained in its right order.

An <u>ambigue</u> I-ideal θ is one whose left and right orders are equal to I.

<u>Proposition 1.</u> *For each prime* p *dividing* DH *there exists exactly one ambigue integral (prime) ideal* P *with the property* $P^2 = I_p$.

All ambigue I*-ideals form an abelian group, and each such ideal has the form*

$$\theta = P_1^{\mu_1} \ldots P_r^{\mu_r}\, m, \quad \mu_1, \ldots = 0 \text{ or } 1$$

where m *is a rational number and* r *is the number of prime factors of* DH.

<u>Corollary</u>. *For* $p \mid H$, *the* p*-component of the corresponding ideal* P *is*

$$P_p = I_p \begin{pmatrix} 0 & 1 \\ p & 0 \end{pmatrix},$$

if I_p *is identified with the order of all* $\begin{pmatrix} a & b \\ pc & d \end{pmatrix}$ *with* $a, \ldots \in \mathbb{Z}_p$.

The <u>proof</u> needs only be carried out for the local components. Indeed, if θ is an ambigue I-ideal, θ_p is an ambigue I_p-ideal. Conversely, if θ_p is a collection of ambigue I_p-ideals which are $= I_p$ for almost all p, the intersection (1) of them is an ambigue I-ideal.

For a p dividing D, the proposition has already been stated above.

Now assume that $p \nmid DH$. Then $I_p = M(2, \mathbb{Z}_p)$. Let $I_p M_p$ be an ambigue I_p-ideal. Because of the elementary divisor theorem M_p may be written

$$M_p = U\begin{pmatrix} p^\alpha & 0 \\ 0 & p^\delta \end{pmatrix} V = U\,D\,V$$

with units U, V and $\alpha < \delta$. The assumption implies that $I_p = M_p^{-1}\,I_p M_p$, and since U and V are units, also $I_p = D^{-1} I_p D$. This is only possible if $\alpha = \delta$, and then $\theta_p = I_p p^\alpha$.

At last we let $p \mid H$. In this situation we employ the following generalization of the elementary divisor theorem:

<u>Lemma</u>. A non-singular matrix M_p can be written in the form

$$(6) \qquad M_p = U\,P^\mu\,D\,V, \quad P = \begin{pmatrix} 0 & 1 \\ p & 0 \end{pmatrix},$$

with $\mu = 0$ or 1 and units U and V of I_p.

With this lemma the proof of proposition 1 proceeds similarly to the case $I_p = M(2,\mathbb{Z}_2)$. It has only to be verified that $P^{-1} I_p P = I_p$. This means that $\theta_p = I_p P$ is an ambigue ideal.

<u>Proof of the lemma</u>. Let $M_p = \begin{pmatrix} a & b \\ pc & d \end{pmatrix} m_p$ with $a, \ldots \in \mathbb{Z}_p$ be given. The factor m_p has no effect and can be omitted.

We either assume that a is less often divisible by p than c, or we multiply M_p on the left by P; then $P\,M_p$ has this property. Let $a = p^\nu a'$, $c = p^\nu c'$ with $(a', c', p) = 1$ and take $\alpha, \beta \in \mathbb{Z}_p$ such that $\alpha a' + p\beta c' = 1$. Then $U = \begin{pmatrix} \alpha & \beta \\ -pc' & a' \end{pmatrix}$ is a unit in I_p and

$$U\,P^\mu M_p = \begin{pmatrix} A & B \\ 0 & D \end{pmatrix} = M_p', \qquad \mu = 0 \text{ or } 1$$

Multiplying M_p' on the left and right by units $U, V = \begin{pmatrix} 1 & t \\ 0 & 1 \end{pmatrix}$ we can reduce B mod A and D. So we can achieve $U\,P^\mu M_p V = \begin{pmatrix} A & 0 \\ 0 & D \end{pmatrix}$, unless A and D are both divisible by p while B is not. In this case

$$M_p'\,P = \begin{pmatrix} pB & A \\ pD & 0 \end{pmatrix} = \begin{pmatrix} B & p^{-1}A \\ D & 0 \end{pmatrix} p = M_p'' p,$$

and M_p'' has a form which was already treated above. At last we observe once more that P generates an ambigue ideal.

§2 <u>Class number and units</u>

Two I-left (right) ideals M_1, M_2 are called left (right) equivalent

Eich-20

if $M_2 = M_1M$ (= MM_1) with an $M \in \Phi$. The class numbers of left and right ideals are equal for all orders of invariant H. For indefinite quaternion algebras it is always 1, but we are here only interested in definite I when the class number is in general $h > 1$.

The analytic method of Dirichlet which yields the class number for algebraic number fields leaves us here with a certain substitute. For a definite quaternion algebra, an order has a finite number of units. Let $I = I_1$ be an order (of invariant H) and $M_1 = I_1, M_2, \ldots, M_h$ a system of I-left ideals representing all classes, and let I_i be the right orders of the M_i and finally e_i the number of units of I_i.

The zeta function of I is the sum

$$\zeta(s) = \sum n(M)^{-2s},$$

summed over all integral I-left ideals M. It can be expressed as the Euler product

$$\zeta(s) = \prod_p \zeta_p(s)$$

where

$$\zeta_p(s) = \sum n(M_p)^{-2s}$$

is summed over all integral M_p whose norms are powers of the prime p. The determination of the number of different M_p whose norm is a given power of p is a simple task of local arithmetic. We have

$$\zeta_p(s) = (1 - p^{-2s})^{-1}(1 - p^{1-2s})^{-1} \qquad \text{for } p\nmid DH,$$
$$\zeta_p(s) = (1 - p^{-2s})^{-1} \qquad \text{for } p|D,$$
$$\zeta_p(s) = (1 + p^{1-2s})(1 - p^{-2s})^{-1}(1 - p^{1-2s})^{-1} \quad \text{for } p|H.$$

Collecting the local contributions we get

$$\zeta(s) = \zeta_Q(2s)\, \zeta_Q(2s-1) \prod_{p_1|D}(1 - p_1^{1-2s}) \prod_{p_2|H}(1 + p_2^{1-2s})$$

with the Riemann zeta function $\zeta_Q(s)$.

On the other hand the zeta functions for the classes are formed:

$$\zeta_i(s) = \sum{}' n(M)^{-2s},$$

summed only over the integral M which are left equivalent with M_i. By definition these M are

$$M = M_i \, M, \qquad M \in M_i^{-1}$$

and

$$n(M) = \frac{1}{n(M_i^{-1})} \, n(M).$$

Let $B_1, \ldots, B_4$ be a basis of M_i^{-1}. Then M runs over all

$$M = m_1 B_1 + \ldots + m_4 B_4, \qquad m_\nu \in \mathbf{Z}$$

except $m_1 = \ldots = 0$. But in this way we count each M e_i times, since M and M U with a unit U yield the same ideal M. Therefore

$$\zeta_i(s) = \frac{1}{e_i} \sum F_i(m_1, \ldots, m_4)^{-2s}$$

where

$$F_i(m_1, \ldots, m_4) = \frac{1}{n(M_i^{-1})} \, n(m_1 B_1 + \ldots + m_4 B_4)$$

is a quadratic form of determinant $D^2 H^2$.

Each $\zeta_i(s)$ has a pole of order 1 at $s = 1$ with the residue $e_i^{-1}(4\pi^2 D\, H)^{-1}$. Observing $\zeta(s) = \sum \zeta_i(s)$ we obtain the measure (of the class number)

(7) $$M = \sum_{i=1}^{h} \frac{1}{e_i} = \frac{1}{24} \prod_{p_1 | D} (p_1 - 1) \prod_{p_2 | H} (p_2 + 1).$$

The class number h will turn up automatically at a later occasion.

§3 Imbedded commutative orders

We are going to study commutative subfields

$$K = Q(M)$$

of Φ where M has trace, norm, and discriminant

Eich-22

$$s(M) = s,\ n(M) = n, \qquad \Delta = s^2 - 4n$$

which is assumed not a square and later $\Delta < 0$. An order $\mathfrak{v}$ of K has always a basis 1, M with respect to $\mathbb{Z}$, and it is uniquely determined by the discriminant Δ.

Ideals for $\mathfrak{v}$ are defined as $\mathbb{Z}$-modules $\mathfrak{m}$ satisfying

$$\mathfrak{v}\,\mathfrak{m} = \mathfrak{m}$$

and for all primes p

$\mathfrak{m}_p = \mathfrak{v}_p M_p$ with $M_p \in K$ and $M_p, M_p^{-1} \in \mathfrak{v}_p$ for almost all p.

The $\mathfrak{v}$-ideals form a multiplicative group. For maximal $\mathfrak{v}$ the second condition is superfluous and can be proved. Ideal classes are defined in the usual way.

Now we assume M contained in a given order I, and then $\mathfrak{v} = \mathbb{Z}(1, M) \subset I$. Let

$$\mathfrak{v}_1 = \mathbb{Z}(1, M_1) = I \cap K.$$

Then $M = u + f\,M_1$ with a natural integer f, and the discriminant of M_1 is

$$\Delta_1 = \Delta f^{-2}.$$

Therefore $\mathfrak{v}_1$ can be only larger than $\mathfrak{v}$ if $\Delta f^{-1} \equiv 0$ or 1 mod 4 with an $f > 1$.

If $\mathfrak{v} = \mathfrak{v}_1$, $\mathfrak{v}$ is called <u>optimally imbedded</u> in I.

We introduce the following modification of the Legendre symbol $(\frac{\Delta}{p})$:

$$\{\frac{\Delta}{p}\} = \begin{cases} 1 \text{ if } \Delta\, p^{-2} \equiv 0 \text{ or } 1 \bmod 4, \\ (\frac{\Delta}{p}) \text{ otherwise,} \end{cases}$$

and we prove

<u>Proposition 2.</u> *An order $\mathfrak{v}$ of discriminant Δ can be imbedded optimally in some order I of invariant* H, *if and only if*

$$\{\tfrac{\Delta}{p}\} \neq 1 \textit{ for all } p_1|D \quad \textit{and } \{\tfrac{\Delta}{p}\} \neq -1 \textit{ for all } p_2|H.$$

Proof. We show this at first locally.

Let $p|D$. K is a splitting field of Φ, and this is the case if and only if $(\frac{\Delta}{p}) \neq 1$. If ν_p is not maximal but contained in the maximal order ν_{1p}, Δp^{-2} is $\equiv 0$ or 1 mod 4 and by definition $\{\frac{\Delta}{p}\} = 1$. Conversely, if $\{\frac{\Delta}{p}\} = 1$, we have $\Delta p^{-2} \equiv 0$ or 1 mod 4, because $(\frac{\Delta}{p}) = 1$ is impossible. So ν is contained in a larger order, and this is contained in I_p, because I_p contains all integral elements.

Let $p \nmid D H$. Then $I_p \cong M(2,\mathbb{Z})$ and without loss of generality, equal. Any solution of an integral equation of second degree can be represented by a matrix in I_p, and it is easy to find a representation M_p which is not of the form $u E + f M_{1p}$ with $f > 1$ and $M_{1p} \in I_p$. So for these p there are no conditions on M_p for being optimally imbedded.

Let $p|H$ and $I_p = \{\begin{pmatrix} a & b \\ pc & d \end{pmatrix}: a,\dots \in \mathbb{Z}_p\}$. If M_p is such a matrix, the discriminant of M_p is not a quadratic non-residue for p. This is $\{\frac{\Delta}{p}\} \neq -1$. Conversely, let this be so. Then M_p can be represented by such a matrix, and with the additional property that $b \not\equiv 0 \bmod p$. Now $\nu_p = \mathbb{Z}_p(E, M_p)$ is optimally imbedded in I_p.

At last we pass from the local to the global imbedding. We have shown that, for each p, there exists an I_p which contains an order isomorphic to ν_p optimally, and I_p is the p-component of an order of invariant H. We can assume that these I_p contain all the same ν_p. If K is imbedded in Φ, which is the case under the assumptions on the $\{\frac{\Delta}{p_1}\} \neq 1$ for $p_1|D$, ν is contained in some maximal order I_o; and if p does not divide the discriminant Δ and also not H, I_{op} contains ν_p optimally and is the p-component of an order of invariant H. This is the case for almost all p. Now the intersection of all I_p exists. It contains ν optimally and has invariant H.

Eich-24

Proposition 3. *Let ν be optimally imbedded in the left and right orders I and I' of an ideal M. Then there exists an ambigue I-ideal Θ and an ν-ideal m such that*

$$M = \Theta\, m. \tag{8}$$

Conversely, if (8) holds, and if ν is contained optimally in I, it is also contained optimally in the right order I' of M.

Proof. The converse part is very easy and will be omitted. For the direct part we show first the existence of an ambigue I_p-ideal Θ_p and ν_p-ideal m_p such that

$$M_p = \Theta_p\, m_p \tag{9}$$

holds.

For $p|D$, (9) is trivial since all ideals are ambigue.

For $p \nmid D\,H$ we assume $I_p = M(2,\mathbf{Z})$ and

$$M_p = I_p M, \quad M = U\, D\, V$$

with unimodular matrices U, V and a diagonal matrix D. The latter can be assumed without loss of generality as $D = \begin{pmatrix} 1 & 0 \\ 0 & p^{\nu} \end{pmatrix}$. The factor U on the left does not effect the ideal M_p and can be omitted. Furthermore one can replace M_p, M, and ν_p by $M_p V^{-1}$, $M\, V^{-1}$, and $\nu_p' = V \nu_p V^{-1}$. ν_p' is optimally imbedded in I_p and the right order $I_p'' = V I_p' V^{-1}$ of $M_p V^{-1}$, and the situation is the same. Therefore we may assume without loss of generality that $M = D = \begin{pmatrix} 1 & 0 \\ 0 & p^{\nu} \end{pmatrix}$.

Let E, $B = \begin{pmatrix} a & b \\ c & d \end{pmatrix}$ be a $\mathbf{Z}$-basis of ν. We can replace B by B - dE so that d becomes 0. The right order of $I_p D$ is $I_p' = D^{-1} I_p D$. The assumption that $B = \begin{pmatrix} a & b \\ c & 0 \end{pmatrix}$ is optimally imbedded in I_p and I_p' means that B and $B' = D\, B\, D^{-1} = \begin{pmatrix} a & p^{-\nu} b \\ p^{\nu} c & 0 \end{pmatrix}$ are optimally imbedded in I_p. This implies at first $b \equiv 0 \bmod p^{\nu}$ and furthermore

either $a \equiv 0 \bmod p$, and both c and $b\, p^{-\nu} \not\equiv 0 \bmod p$,

or $a \not\equiv 0 \bmod p$.

In the first case $B\,D^{-1} = W$, in the second $(B + \rho p^{\nu}E)\,D^{-1} = W$ with a certain ρ, is a unit. Thus (9) is satisfied with either $m_p = \nu_p B$ or $\nu_p(B + \rho p^{\nu}E)$.

At last let $p|H$. Now we assume $M_p = I_p M$ with M in the form (6). If the factor P occurs, it generates an ambigue ideal as allowed in (9). Dividing by P we need only treat the case when $M = U\,D\,V$ with unimodular matrices U, V and a diagonal matrix $D = \begin{pmatrix} 1 & 0 \\ 0 & p^{\nu} \end{pmatrix}$. As before, we may assume without loss of generality $M = D$.

$$B = \begin{pmatrix} a & b \\ pc & 0 \end{pmatrix} \text{ and } B' = D\,B\,D^{-1} = \begin{pmatrix} a & p^{-\nu}b \\ p^{\nu+1}c & 0 \end{pmatrix}$$

are optimally imbedded in I_p. Therefore $b \equiv 0 \bmod p^{\nu}$ and

either $a \equiv 0 \bmod p$, and both c and $bp^{-\nu} \not\equiv 0 \bmod p$,

or $a \not\equiv 0 \bmod p$.

In the first case $B\,D^{-1}P^{-1} = W$, and in the second $(B + p^{\nu}E)\,D^{-1}P^{-1} = W$ is a unit. Thus (9) holds with $\theta_p = I_p P$ and $m_p = \nu_p M$ or $\nu_p(M + p^{\nu}E)$.

The local constructions of the θ_p and m_p being achieved, we put

$$\theta = \cap\theta_p, \quad m = \cap m_p.$$

These intersections exist because $\theta_p = I_p$ and $m_p = \nu_p$ for almost all p. These ideals satisfy (8).

§4 The classes of elements in an order

In this section we assume $I = \left(\begin{pmatrix} a & b \\ Hc & d \end{pmatrix} : a, \ldots \in \mathbb{Z}\right)$. It is easy to show that all I-ideals are principle. Indeed let $M_p = I_p M_p$ and $|M_p| = u_p p^{\nu_p}$ with p-adic units u_p. Without loss of generality one can assume $M_p \in I_p$ and the product of all those u_p for which $\nu_p \neq 0$, is 1. Now choose an $M \in I$ satisfying the (finitely many) congruences

Eich-26

$$M \equiv M_p \bmod I_p p^{\nu_p}$$

and with the determinant $|M| = \Pi p^{\nu_p}$. This is possible by virtue of the assumption on the u_p. With this M we have $M = IM$.

We shall call two elements A, A_1 equivalent, if $A_1 = U\,A\,U^{-1}$ with a unit U of I. Furthermore we say that A is optimally imbedded in I, if the order $\nu = \mathbb{Z}(1, A)$ is optimally imbedded.

Proposition 4. *The number of classes of optimally imbedded elements* A *with discriminant* Δ *(not a square) is*

$$\Pi_{p|H}\,(1 + \{\tfrac{\Delta}{p}\})\,h(\Delta),$$

where $h(\Delta)$ *is the number of ideal classes of the order* $\nu = \mathbb{Z}(1, A)$.

These classes can be gathered in families which contain with an A *all* $A_1 = M\,A\,M^{-1}$ *for which* $IM = Im$ *with an* ν*-ideal* m. *Each family contains* $h(\Delta)$ *classes.*

If A *and* $A_1 = M\,A\,M^{-1}$ *are optimally imbedded in* I*, but not in the same family,* $IM = \theta m$ *with an* ν*-ideal* m *and an ambigue* I*-ideal* θ *which is not of the form* $\theta = Im'$ *with an* ν*-ideal* m'.

Proof. If A and A_1 have the same traces and norms, there exists an M such that $A_1 = M\,A\,M^{-1}$. Let both be optimally imbedded in I. Then ν is also optimally imbedded in $M^{-1}IM$ which is the right order of IM. By proposition 3,

$$IM = \theta\, m$$

with an ambigue ideal θ and an ν-ideal m. Proposition 1 allows us to assume

$$IM = P_1^{\mu_1} \ldots P_r^{\mu_r} m \qquad \mu_i = 0 \text{ or } 1 \tag{10}$$

with the ambigue prime ideals P_i corresponding to the prime factors p_i of H.

Some of the P_i may be of the form $P = I_p$ with an ν-prime ideal p.

We contend that this is the case if and only if $\{\frac{\Delta}{p}\} = 0$. Indeed, from $P^2 = Ip$ follows $p^2 = \nu p$, and p is an ambigue ν-ideal. Such exist only if ν_p is maximal and $\Delta \equiv 0 \bmod p$, in other words if $\{\frac{\Delta}{p}\} = 0$. Conversely, if $\{\frac{\Delta}{p}\} = 0$, then there exists a p with $p^2 = \nu p$. Let $p_p = \nu_p\begin{pmatrix} a & b \\ pc & d \end{pmatrix}$. Then $a \equiv d \equiv 0 \bmod p$, and $I_p p_p$ is divisible by $I_p\begin{pmatrix} 0 & 1 \\ p & 0 \end{pmatrix}$ which implies $P = I_p$.

We include as many of the $P_i^{\mu_i}$ in m as possible. Then there remain exactly $\Pi(1 + \{\frac{\Delta}{p}\})$ possibilities for the remaining factor $P_1^{\mu_1}\ldots$.

If M is multiplied on the left by a unit U, the ideal (10) does not change, and A_1 is replaced by $U A_1 U^{-1}$, an equivalent element.

Let $M_1 \in K$ and replace M and m in (10) by $M' = M M_1$ and $m' = mM_1$. ν is again optimally imbedded in the order $I' = M'^{-1} I M'$, and $\nu_1' = M'\nu M'^{-1}$ is optimally contained in I. Because $M_1 \in K$, $\nu_1' = \nu_1$, and also A_1 has not changed.

So we have seen: to a class of elements A_1 optimally imbedded in I, there corresponds a pair, consisting of an ambigue ideal $\theta = P_1^{\mu_1}\ldots$ with $\mu_i = 0$ or 1 and such that no $P^\mu = Ip$ with an ν-ideal p occurs with $\mu = 1$, and of an ν-ideal class.

The converse correspondence also holds. Let θ and m be such ideals. The product (10) is a principle ideal, and $A_1 = M A M^{-1}$ is optimally contained in I, because A is optimally contained in the right order of IM.

§5 <u>The classes of elements in an order, continued</u>

In this section we assume Φ as definite. The situation is now rather different. The number of ideal classes is in general > 1. As in § 2 we select an order $I = I_1$ of invariant H and a system of I-left ideals $M_1 = I_1, \ldots, M_h$ representing all left classes. The right orders of the M_i will be denoted by I_i, and the numbers of units of I_i by e_i. (The I_i are not necessarily different).

Let an element A with integral trace, norm, and discriminant be given:

Eich-28

$$s(A) = s, \quad n(A) = n, \qquad \Delta = s^2 - 4n.$$

If

$$\{\frac{\Delta}{p_1}\} \neq 1 \text{ for all } p_1|D \text{ and } \quad \{\frac{\Delta}{p_2}\} \neq -1 \text{ for all } p_2|H$$

there exists an order I of invariant H which contains A optimally (proposition 2). We assume that $I = I_1$ has this property.

This time the number of elements A_i with this trace and norm which are optimally imbedded in I_i is finite. Evidently this number depends only on Δ; we denote it by $a_i(\Delta)$. With a unit U_i of I_i, also $A_i' = U_i A_i U_i^{-1}$ is optimally imbedded in I_i, and $A_i' \neq A_i$ unless $U_i \in \nu_i = \mathbb{Z}(1,A_i)$. If $w(\Delta)$ is the number of units of $\nu = \mathbb{Z}(1,A)$, the $A_i \in I_i$ form $a_i(\Delta)\ \frac{w(\Delta)}{e_i}$ classes of equivalent elements, each class containing $\frac{e_i}{w(\Delta)}$ elements.

With these notations and assumptions we propose to prove ($h(\Delta)$ is again the ideal class number of $\nu = \mathbb{Z}(1, A)$).

<u>Proposition 5.</u> *The numbers of classes of equivalent* A_i *with trace* s *and norm* n *which are optimally imbedded in* I_i, *summed over all* i, *is*

$$\sum_{i=1}^{h} a_i(\Delta)\ \frac{w(\Delta)}{e_i} = \Pi_{p_1|D}(1 - \{\frac{\Delta}{p_1}\})\ \Pi_{p_2|H}(1 + \{\frac{\Delta}{p_2}\})\ h(\Delta).$$

These classes can be gathered in families, a pair $A_i \in I_i$ *and* $A_j \in I_j$ *belonging to the same family if and only if* $A_j = M\ A_i M^{-1}$ *and*

$$M_i^{-1}\ M_j\ M = I_i\ m_i$$

with an ν_i*-ideal* m_i. *In each family there are* $h(\Delta)$ *classes.*

The <u>proof</u> is very similar to that of proposition 4. Let $A \in I = I_1$ and $A_i \in I_i$ be optimally imbedded. Since $A_i = M\ A\ M^{-1}$ with some M, A is optimally imbedded in $M^{-1}\ I_i\ M$ which is the right order of $M_i M$. Now we have

$$(11) \qquad M_i\ M = P_1^{\mu_1} \ldots P_r^{\mu_r}\ m, \qquad \mu_\nu = 0 \text{ or } 1$$

where the P_ν are the ambigue prime ideals belonging to the prime factors p_ν of D H. If we multiply M on the left by a unit U_i of I_i we get an element equivalent with A_i, and if we multiply M on the right by an element $M_1 \in K$, A_i does not change. Thus to the classes of equivalent A_i lying in one of the I_i there correspond pairs of an ambigue ideal $\theta = P_1^{\mu_1}\ldots$ no factor of which can be expressed in the form $P = Ip$, and of a class of ν-ideals.

Conversely, if such a pair is given, the product on the right of (11) is left equivalent to one of the M_i, and $M A M^{-1}$ is contained optimally in I_i because A is contained optimally in the right order of $M_i M$.

§6 The Brandt matrices

A few preparations are necessary. Again let Φ be definite and consider two generating elements Ω_1, Ω_2 with

$$\Omega_1^2 = -p_1, \quad \Omega_2^2 = -p_2, \qquad \Omega_1\Omega_2 + \Omega_2\Omega_1 = 0$$

and positive $p_1, p_2 \in Q$. We can write

$$\Omega_1 = \sqrt{p_1}\, I_1, \quad \Omega_2 = \sqrt{p_2}\, I_2, \qquad I_1 I_2 + I_2 I_1 = 0,$$

where I_1, I_2 and $I_3 = I_1 I_2$ are Hamilton quaternions. The general element of Φ can therefore be written

$$Z = x_0 + x_1 I_1 + x_2 I_2 + x_3 I_3$$

with real $x_0, \ldots$. The Hamilton quaternions, on the other hand can be represented by the complex matrices

$$I_1 = \begin{pmatrix} i & 0 \\ 0 & -i \end{pmatrix}, \quad I_2 = \begin{pmatrix} 0 & 1 \\ -1 & 0 \end{pmatrix}, \quad I_3 = \begin{pmatrix} 0 & i \\ i & 0 \end{pmatrix}.$$

Therefore the multiplicative group Φ^* allows a representation

$$(12) \qquad Z \to X_1(Z) = \begin{pmatrix} z_1 & z_2 \\ -\bar{z}_2 & \bar{z}_1 \end{pmatrix}, \quad \begin{matrix} z_1 = x_0 + ix_1, & z_2 = x_2 + ix_3, \\ \bar{z}_1 = x_0 - ix_1, & \bar{z}_2 = x_2 - ix_3. \end{matrix}$$

Let $\begin{pmatrix} \eta_0 \\ \eta_1 \end{pmatrix}$ be the basis of the representation module, written as a

Eich-30

column vector. The representation is defined by

$$\begin{pmatrix}\eta_0\\ \eta_1\end{pmatrix} Z = \begin{pmatrix}\eta_0 Z\\ \eta_1 Z\end{pmatrix} = X_1(Z)\begin{pmatrix}\eta_0\\ \eta_1\end{pmatrix}.$$

Out of this we form the (l+1)-rowed representation

$$(13)\qquad \begin{pmatrix}\eta_0^l\\ \eta_0^{l-1}\ \eta_1\\ \vdots\\ \eta_1^l\end{pmatrix} Z = \begin{pmatrix}(\eta_0 Z)^l\\ (\eta_0 Z)^{l-1}\ (\eta_1 Z)\\ \vdots\\ (\eta_1 Z)^{l+1}\end{pmatrix} = X_l(Z)\begin{pmatrix}\eta_0^l\\ \eta_0^{l-1}\ \eta_1\\ \vdots\\ \eta_1^l\end{pmatrix}.$$

We also include the identical representation $X_0(Z) = 1$.

Proposition 6. *The elements of the matrix* $X_l(Z)$ *are homogeneous polynomials* $p_l(x_0, \dots, x_3)$ *satisfying the Laplace differential equation*

$$\Delta\, p_l(x) = 4\left(\frac{\partial^2}{\partial z_1 \partial \bar{z}_1} + \frac{\partial^2}{\partial z_2 \partial \bar{z}_2}\right) p_l(x) = 0$$

For the proof it suffices to show that the monomials

$$(\eta_0 Z)^\lambda (\eta_1 Z)^{l-\lambda} = (z_1\eta_0 + z_2\eta_1)^\lambda (-z_2\eta_0 + z_1\eta_1)^{l-\lambda}$$

with the indeterminates η_0, η_1 are solutions of the Laplace equation.

We also note

Proposition 7. *For an element* $A \in \Phi$ *with trace* s *and norm* n, *the trace of the representing matrix is*

$$s(X_l(A)) = \frac{\rho^{l+1} - \bar{\rho}^{l+1}}{\rho - \bar{\rho}} \qquad \textit{where } \rho = \tfrac{1}{2}(s + \sqrt{s^2 - 4n})$$

and $\bar{\rho}$ *is the complex conjugate.*

For the proof one transforms A by a complex matrix into $\begin{pmatrix}\rho & 0\\ 0 & \bar{\rho}\end{pmatrix}$ which leaves the trace of $X_l(A)$ unchanged.

Now let the M_i and I_i have the same meaning as in § 5. We consider all integral ideals N_{ij} of a given norm n with left order I_i which are left equivalent with $M_i^{-1}M_j$. They can be written in the form

$$(14)\qquad N_{ij} = M_i^{-1}\, M_j\, A_{ij}.$$

With the transposed matrices of the $X_l(Z)$ we define the <u>Brandt matrices</u> by

(15) $$B_l(n) = B_l(n;\ D,\ H) = (\sum X_l^t(A_{ij})\ e_j^{-1}).$$

They are h-rowed matrices, at the index ij stand the sums $\sum X_l^t(A_{ij})\ e_j^{-1}$, summed over all A_{ij} such that (14) is an integral ideal of norm n. So the $B_l(n)$ have eventually $(l+1)h$ rows. The definition is supplemented by

(15a) $$B_0(0) = \begin{pmatrix} e_1^{-1} & \cdots & e_h^{-1} \\ \hdashline e_1^{-1} & \cdots & e_h^{-1} \end{pmatrix}, \quad B_l(0) = 0 \text{ for } l > 0.$$

<u>Theorem 1.</u> *The (coefficients of the matrix) series*

(16) $$\theta_l(z;\ D,\ H) = \sum_{n=0}^{\infty} B_l(n;\ D,\ H)\ e^{2\pi i\ n\ z}$$

are modular forms with respect to the congruence subgroup $\Gamma_0(N)$ *with* $N = D\ H$, *of weight* $l+2$, *and of character* 1.

<u>Proof</u>. Fix a pair of indices ij, and take a basis $B_1,\dots,B_4$ of the ideal $M_j^{-1}\ M_i$. Then the elements A_{ij} in (14) can be written in the form

$$A_{ij} = \sum m_\nu\ B_\nu$$

When the m_ν run over $\mathbb{Z}^4$, (14) runs over all integral N_{ij}, each occurring e_j-times. Thus

$$\sum_{n=0}^{\infty} (\sum X_l^t(A_{ij})\ e_j^{-1})\ e^{2\pi inz} = \theta_{ij}(z)$$

is the theta function attached to the quadratic form

$$F_{ij}[m] = \frac{n(M_j)}{n(M_i)}\ n(\sum m_\nu B_\nu)$$

and the polynomials $X_l^t(\sum m_\nu B_\nu)\ e_j^{-1}$ by proposition 6, are of the form considered in §1.

Eich-32

<u>Theorem 2.</u> *The Brandt matrices have the following properties:* $(n > 0)$

$$(17)\qquad B_l(n)^t = \begin{pmatrix} E_{l+1}n(M_1)^{-l}e_1 & & \\ & \ddots & \\ & & E_{l+1}n(M_h)^{-l}e_h \end{pmatrix}^{-1} B_l(n)\ .$$

$$\cdot \begin{pmatrix} E_{l+1}n(M_1)^{-l}e_1 & & \\ & \ddots & \\ & & E_{l+1}n(M_h)^{l}e_h \end{pmatrix}$$

$$(18)\qquad B_l(n_1)\ B_l(n_2) = B_l(n_1n_2) \quad \text{for } (n_1, n_2) = 1,$$

$$(19)\quad B_l(p^{\mu})\ B_l(p^{\nu}) = \sum_{\sigma=0}^{\min(\mu,\nu)} p^{(l+1)\sigma} B_l(p^{\mu+\nu-2\sigma}) \text{ for primes } p \nmid D\ H,$$

$$(20)\qquad B_l(p^{\mu})\ B_l(p^{\nu}) = B_l(p^{\mu+\nu}) \quad \text{for primes } p \mid D.$$

They generate a semisimple commutative ring.

<u>Proof</u>. The semisimplicity follows from the commutativity, namely (18), and the fact that the matrices $\left(E_{l+1}\sqrt{n(M_i)^{-l}e_i}\right)^{-1} B_l(n) \left(E_{l+1}\sqrt{n(M_i)^{-l}e_i}\right)$ are Hermitean (see (17)).

Proof of (17). Let $n > 0$.

From the definition (12) and (13) we take $\overline{X}_l^t(Z) = X_l(Z')$ where Z' is the conjugate quaternion to Z. Hence the right hand side of (17) is

$$(21)\qquad \frac{n(M_i)^l}{n(M_j)^l} \sum \overline{X}_l\ (A'_{ij})\ e_i^{-1}\ e_j$$

The ideals conjugate to the N_{ij} are

$$N'_{ij} = A'_{ij}\ M'_j\ M_i'^{-1} = A'_{ij}\ (M_j^{-1}\ M_i\ B_{ji})\ A_{ij}'^{-1}$$

with

$$(22)\qquad B_{ij} = \frac{n(M_j)}{n(M_i)}\ A'_{ij}.$$

They run over all integral I_i-right ideals of norm n which are right equivalent with $M_j^{-1} M_i$. And the

$$N''_{ji} = M_j^{-1} M_i B_{ji}$$

run over all integral I_j-left ideals which are left equivalent with $M_j^{-1} M_i$. The N''_{ji} represent these e_i-times each. Now we get from (21), and (22)

$$\frac{n(M_i)^l}{n(M_j)^l} \sum \overline{X}_l (A'_{ij}) e_i^{-1} = \sum \overline{X}_l (B_{ji}) e_i^{-1}$$

which is (17).

Proof of (18). An ideal N_{ik} of norm $n_1 n_2$ with relatively prime n_1, n_2 is uniquely decomposable into a product

$$N_{ik} = N_{ij} N_{jk} \tag{23}$$

of integral ideals of norms n_1, n_2. We express both factors in the way (14)

$$N_{ik} = M_i^{-1} M_k A_{ik} = M_i^{-1} M_j B_{ij} \cdot B_{ij}^{-1} M_j^{-1} M_k C_{jk} B_{ij}$$

whence

$$A_{ik} = C_{jk} B_{ij}. \tag{24}$$

$\tilde{N}_{jk} = M_j^{-1} M_k C_{jk}$ is an integral ideal of norm n_2 and of left order I_j and left class $M_j^{-1} M_k$. Let B_{ij} and C_{jk} run over all elements such that N_{ij} and $\tilde{N}_{jk}$ are integral ideals of norms n_1, n_2, each occurring e_j-resp. e_k-times. Then the product (23) with $N_{jk} = B_{jk}^{-1} \tilde{N}_{jk} B_{ij}$ yields all integral ideals of norm $n_1 n_2$. In this way, each product N_{ik} occurs $e_j e_k$-times. Therefore each element A_{ik} in (24) occurs e_j-times, and from (24) follows

$$\sum_j (\sum X_l^t(B_{ij}) e_j^{-1} \cdot \sum X_l^t(C_{jk}) e_k^{-1}) = \sum \sum X_l^t(C_{jk} B_{ij})(e_j e_k)^{-1} = \sum X_l^t(A_{ik}) e_k^{-1}$$

which is (18).

Eich-34

Equation (19) we prove at first for $\nu = 1$, and for "reduced" Brandt matrices $B_1^0(n)$, counting only primitive ideals N_{ij}. A primitive integral ideal N_{ik} of norm $p^{\mu+1}$ is also uniquely decomposable into the product (23) with primitive integral factors of norms p^μ and p. But if N_{ij} and N_{jk} have this property, the product (23) need not be primitive. It can have the form $N_{ik} = p\tilde{N}_{ik}$ with a primitive $\tilde{N}_{ik}$, and this is the case if and only if N_{ij} is divisable on the right by the ideal N'_{jk}, conjugate to N_{jk}. Thus, when the N_{ij}, N_{jk} run over all primitive integrals of these kinds, the products (23) yield all primitive N_{ik} once, and all $p\tilde{N}_{ik}$ with primitive $\tilde{N}_{ik}$ as often as integral N_{ij} of norm p exist, i.e. (p+1)-times. This can be expressed as

$$(25) \qquad B_1^0(p^\mu)\; B_1(p) = B_1^0(p^{\mu+1}) + p^1(p+1)\; B_1^0(p^{\mu-1}).$$

The original Brandt matrices are

$$B_1(p^\mu) = \textstyle\sum_{\nu \geq 0} p^1\; B_1^0(p^{\mu-2\nu}),$$

and with this (25) becomes

$$(26) \qquad B_1(p^\mu)\; B_1(p) = B_1(p^{\mu+1}) + p^{1+1}\; B_1(p^{\mu-1}).$$

From (26), (19) can be derived by a simple inductive argument which was first used by Hecke in a similar situation.

We add two corollaries:

Corollary 1. *The Brandt matrices of index* 0 *can be simultaneously reduced to*

$$B_0(n) = \begin{pmatrix} B_0(n)' & \\ & b(n) \end{pmatrix},$$

where b(n) *is a one-rowed component, equal to the number of integral left (or right) ideals of norm* n, *i.e. the coefficient in the zeta function*

$$\zeta(s) = \textstyle\sum_{n=1}^{\infty} b(n)\; n^{-2s},$$

as given in No. §2.

Proof. The sum over the rows of $B_0(n)\begin{pmatrix} e_1 & & \\ & \ddots & \\ & & e_h \end{pmatrix}$ is = b(n) for every row.

Therefore $B_0(n) \to b(n)$ is a homomorphism, and the corollary follows from the theorem.

The following is immediately clear in the connection of the proof of the theorem:

Corollary 2. *For a prime* p *dividing* H *let* $B_1(n, p^\nu; D, H)$ *be the sum (15) in which the* A_{ij} *are restricted to such ideals (14) which are divisible by the ambigue prime ideal* P_i *exactly* ν *times. Then*

$$B_1(p, p; D, H)\ B_1(n, p^\nu; D, H) = B_1(pn, p^{\nu+1}; D, H).$$

§7 The Brandt matrices with a character

The object of this section is the modification of the Brandt matrices which allows to extend (20) to primes dividing H. Now we assume the M_i as integral and with norms which are relatively prime to H. Then we have for all p dividing D or H:

$$(27) \qquad (M_i^{-1} M_j)_p = I_{i\,p} = I_{1\,p} = I_p.$$

For each prime ideal P attached to a $p|D$ the residues of I mod P form a field of p^2 elements, and the elements $\not\equiv 0$ mod P_i form a cyclic group. Let R_i be a generating element and χ_i a character of this group. We extend χ_i to the whole field $I|P_i$ by demanding $\chi_i(0) = 0$. χ_i will then be called a proper character of $I|P_i$. In the case $\chi_i(R_i) = 1$ we can also put $\chi_i(0) = 1$, and now we speak of an improper character.

For a P_i attached to a $p_i|H$ the ring $I|P_i$ is isomorphic with that of the matrices $M = \begin{pmatrix} a & 0 \\ 0 & d \end{pmatrix}$ with a, d $\in \mathbb{Z}/p_i\mathbb{Z}$. In this case we fix a character

$$\chi_i(M) = \chi_i(a)$$

Eich-36

which is 0 for $a \equiv 0 \bmod p_i$ and a character of the multiplicative group $(\mathbb{Z}/p_i\mathbb{Z})^*$. Let a_i be a generating element of this group and $R_i \equiv \begin{pmatrix} a_i & 0 \\ 0 & 1 \end{pmatrix} \bmod P_i$. For these P_i the characters are always proper ones.

Eventually we put

$$(28) \qquad \chi(M) = \Pi \chi_i(M),$$

multiplied over all primes dividing DH. (28) defines a character mod DH, and the conductor is equal to DH if all characters χ_i are proper.

<u>The Brandt matrices with character</u> χ are

$$(29) \qquad B_l(n; D, H, \chi) = \sum \chi(A_{ij})\, X_l^t(A_{ij})\, e_j^{-1}$$

and

$$B_l(0, D, H, \chi) = 0$$

unless H = 1 and all characters are improper ones. But in this case the Brandt matrices are those without a character.

<u>Theorem 3.</u> *The (coefficients of the matrix) series*

$$(31) \qquad \theta_l(z; D, H, \chi) = \sum_{n=0}^{\infty} B_l(n; D, H, \chi)\, e^{2\pi i N^{-1} n z}$$

are modular forms with respect to the congruence subgroup $\Gamma(N)$ *with* $N = DH$, *of weight* $l + 2$. *For all* $G \in \Gamma$ *with* $G \equiv \begin{pmatrix} a & 0 \\ 0 & d \end{pmatrix}$ *mod* N *we have*

$$\theta_l(z; D, H, \chi)\, [G]^{-l-2} = \chi(d)\, \theta_l(z; D, H, \chi).$$

In the <u>proof</u> we apply the same notations as in that of theorem 1. Furthermore we choose an element $R \in I$ which satisfies the congruences

$$R \equiv R_i \bmod P_i$$

for all ambigue prime ideals P_i. We form the series

$$\sum_{n=0}^{\infty} \left(\sum_{R^\rho} X_l^t(A_{ij}) e_j^{-1}\right) e^{2\pi i N^{-1} n z} = \theta_{ij}(z; R^\rho)$$

where $\sum_R \rho$ means the sum over all $A_{ij} = \sum m_\nu B_\nu$ which satisfy the congruences

$$A_{ij} \equiv R^\rho \bmod \Pi_k\, P_k.$$

More correctly, we ought to demand this congruences locally for each prime p_k dividing N; but this is practically the same because of (27).

The situation can be expressed in another way. We remember that $\theta = \Pi\, P_i$ is the different of I. Let B' be a basis of $M_j^{-1} M_i\, \theta_i$. Then

$$A_{ij} = \sum (m_\nu' + N^{-1} r_\nu')\, B_\nu',$$

and m_ν' runs over $\mathbf{Z}^4$ while r_ν' is defined by

$$N^{-1} \sum r_\nu'\, B_\nu' = R_{ij}^\rho \quad \text{and} \quad R_{ij}^\rho \equiv R^\rho \text{ locally mod all } p_i | N = DH.$$

To $M_j^{-1} M_i\, \theta_i$ is attached the primitive quadratic form F_{ij}' with

$$n(\sum m_\nu'\, B_\nu') = n(M_j^{-1} M_i)\, N\, F_{ij}'[m'],$$

and because θ is the different of I_i,

$$F_{ij}'\, r_\nu' \equiv 0 \bmod N.$$

Now we write

$$n = n(M_i^{-1} M_j\, A_{ij}) = N\, F_{ij}'[m' + N^{-1} r'],$$

and (30) assumes the form of the theta series §1, (5). It satisfies the functional equations Ch. I, §2 (19):

$$\theta_{ij}(z, R^\rho)\, [\begin{smallmatrix} a & b \\ c & d \end{smallmatrix}]^{-1-2} = \theta_{ij}(z, d^{-1}R) \text{ for } (\begin{smallmatrix} a & b \\ c & d \end{smallmatrix}) \equiv (\begin{smallmatrix} d^{-1} & 0 \\ 0 & d \end{smallmatrix}) \bmod N.$$

Therefore

$$\theta_{ij}(z;\, D,\, H,\, \chi) = \sum_\rho \chi(R^\rho)\, \theta_{ij}(z, R^\rho)$$

has the property

$$\theta_{ij}(z;\, D,\, H,\, \chi)\, [\begin{smallmatrix} a & b \\ c & d \end{smallmatrix}]^{-1-2} = \chi(d)\, \theta_{ij}(z;\, D,\, H,\, \chi)$$

for $b \equiv c \equiv 0 \bmod N = DH$, q.e.d. .

Eich-38

Theorem 4. *The Brandt matrices with character* χ *have the following properties:*

$$(32)\quad \overline{B_l(n;\, D,\, H,\, \chi)}^t = \begin{pmatrix} E_{l+1}n(M_1)^{-l}e_1 & & \\ & \ddots & \\ & & E_{l+1}n(M_h)^{-l}e_h \end{pmatrix}^{-1} B_l(n;\, D,\, H,\, \overline{\chi}) \cdot$$

$$\cdot \begin{pmatrix} E_{l+1}n(M_1)^{-l}e_1 & & \\ & \ddots & \\ & & E_{l+1}n(M_h)^{-l}e_h \end{pmatrix}$$

$$(33)\quad B_l(n_1;\, D,\, H,\, \chi)\; B_l(n_2;\, D,\, H,\, \chi) = B_l(n_1 n_2;\, D,\, H,\, \chi) \text{ for } (n_1, n_2) = 1,$$

$$(34)\quad B_l(p^\mu; D,H,\chi)\; B_l(p^\nu; D,H,\chi) = \sum_{\sigma=0}^{\min(\mu,\nu)} p^{(l+1)\sigma}\chi(p^{2\sigma})\; B_l(p^{\mu+\nu-2\sigma}; D,H,\chi).$$

This holds for all p *but becomes simple for* $p|N$ *because then* $\chi(p^\sigma) = 0$.

The proof of theorem 2 is practically applicable in this slightly more general situation.

§8 The traces of the Brandt matrices

We treat at first the case of those without a character. Equation (15) and proposition 7 imply

$$s(B_l(n;\, D,\, H)) = \sum_{i=1}^{h} a_i(s,n)\, \frac{\rho^{l+1} - \overline{\rho}^{l+1}}{\rho - \overline{\rho}} \quad \text{with } \rho = \frac{s + \sqrt{s^2 - 4n}}{2}$$

where $a_i(s, n)$ means the number of all $A_{ii} \in I_i$, each counted with multiplicity e_i^{-1}, which have trace s and norm n. These A_{ii} are classified accordingly as $\mathbb{Z}(1, A_{ii})$ is optimally imbedded in I_i or not. More precisely, let

$$\nu_0(A_{ii}) = I_i \cap Q(A_{ii})$$

and $\Delta_0(A_{ii})$ be the discriminant of this order, and

$$s(A_{ii}) = s,\; n(A_{ii}) = n,\quad \Delta(A_{ii}) = s^2 - 4n = \Delta_0(A_{ii})f^2$$

with a natural integer f. Then to each $A_{ii} \in I_i$ we have the pair s, f, where $(s^2 - 4n)\,f^{-2} \equiv 0$ or $1 \bmod 4$, and the order of discriminant $(s^2 - 4n)\,f^{-2}$ is optimally imbedded in I_i. Now we can apply proposition 5 and obtain

Theorem 5. *The trace of the Brandt matrix is*

$$(35)\qquad s_l(n;\, D,\, H) = \sum_{s,f} \frac{\rho^{l+1}-\bar{\rho}^{l+1}}{\rho-\bar{\rho}}\, \frac{h((s^2-4n)f^{-2})}{w((s^2-4n)f^{-2})}\,\cdot$$

$$\cdot\, \prod_{p_1|D}\left(1-\left\{\frac{(s^2-4n)f^{-2}}{p_1}\right\}\right) \prod_{p_2|H}\left(1+\left\{\frac{(s^2-4n)f^{-2}}{p_2}\right\}\right) + \begin{cases} 2(l+1)Mn^{1/2} \\ 0 \end{cases}$$

according as n *is a rational square or not. Here* s, f *run over all integers with*

$$s^2 - 4n < 0,\quad (s^2 - 4n)\, f^{-2} \equiv 0 \text{ or } 1 \bmod 4,\ f > 0$$

and ρ *as in proposition 7.* $h(\Delta)$ *and* $w(\Delta)$ *denote the numbers of ideal classes and units of the order of discriminant* Δ. M *is the measure, given in (7).*

$2(l + 1)\, M\, n^{1/2}$ must be added in the case $\sqrt{n} \equiv 0 \bmod 1$ because then $A_{ii} = \pm\sqrt{n}$ is possible. An application of theorem 5 is

Theorem 6. *The class number of left (or right) ideals of an order of square-free invariant* H *is*

$$(36)\qquad h = s(B_0(1;\, D,\, H)) = \frac{1}{12}\prod_{p_1|D}(p_1-1) \prod_{p_2|H}(p_2+1) +$$

$$+ \frac{1}{4}\prod_{p_1|D}\left(1-\left(\frac{-4}{p_1}\right)\right)\prod_{p_2|H}\left(1+\left(\frac{-4}{p_2}\right)\right) + \frac{1}{3}\prod_{p_1|D}\left(1-\left(\frac{-3}{p_1}\right)\right)\prod_{p_2|H}\left(1+\left(\frac{-3}{p_2}\right)\right).$$

For the Brandt matrices of character χ we must gather together the A_{ii} in the families which are described in proposition 5. A_{ii} and A'_{ii} with the same trace and norm which are both optimally imbedded in I_i are connected by $A'_{ii} = M\, A_{ii} M^{-1}$. For all $p|N$ we have

$$M = U_p\, P^{\mu} M_{1p},\qquad P = \begin{pmatrix} 0 & 1 \\ p & 0 \end{pmatrix},\ \mu = 0 \text{ or } 1$$

Eich-40

with a unit U_p of I_{ip} and $M_{1p} \in Q(A_{ii})$. It is assumed that $I_{ip}P \neq I_{ip}M_{2p}$ with an $M_{2p} \in Q(A_{ii})$.

For a $p|H$ we let $A_{ii} \equiv \binom{a\ 0}{0\ d}$ mod P, then the contribution of this p to the character χ is by definition (in §7) $\chi_p(A_{ii}) = \chi_p(a)$. Now, if $\mu = 1$ for this p, $A'_{ii} \equiv \binom{d\ 0}{0\ a}$ mod P and consequently $\chi_p(A'_{ii}) = \chi_p(d) = \chi_p(\frac{n}{a})$.

For the $p|D$, the characters A_{ii} and $P^{-1}A_{ii}P$ are the same because the residues of I mod P form a commutative field. Thus we have

<u>Theorem 7.</u> *Let* a_p, d_p *be defined by*

$$x^2 - s\,x + n \equiv (x - a_p)(x - d_p) \ mod\ p \quad \text{for } p|H.$$

Then the trace for the Brandt matrix with the character

$$\chi(A) = \Pi_{p_1|D}\, \chi_{p_1}(A)\ \Pi_{p_2|H}\, \chi_{p_2}(A)$$

for an $A \in I$ *is*

$$(37) \qquad s(B_1(n;\, D,\, H,\, \chi)) = \sum\nolimits_{s,f} \frac{\rho^{1+1}-\bar{\rho}^{1+1}}{\rho-\bar{\rho}}\ \Pi_{p_1|D}(1-\{\tfrac{\Delta}{p_1}\})\chi_{p_1}(n)\ .$$

$$\cdot\ \Pi_{p_2|H}(1+\{\tfrac{\Delta}{p_2}\})\ \frac{\chi_{p_2}(a_{p_2})+\chi_{p_2}(d_{p_2})}{2}\ \frac{h(\Delta)}{w(\Delta)} + \begin{cases} 2(1+1)Mn^{1/2}\chi(\sqrt{n}) \\ 0 \end{cases}$$

according as n *is a rational square or not. The sum is extended over all* s, f *as in theorem 5,* Δ *means* $(s^2-4n)f^{-2}$.

§9 <u>Remarks on the literature</u>

As we said in the introduction, we hope that an experienced reader will be able to carry out the proofs for the facts summarized in §1 and in §2, even if he is not yet familiar with the theory of non-commutative algebras. If he wants more information he is referred to <u>Deuring's</u> book [3] for the case of maximal orders and to [4] for our non-maximal ones.

The results on imbedded commutative orders in the case of maximal quaternion orders are due to Chevalley [2], Hasse [6], and Noether [8]. The present author extended them to our non-maximal orders [4].

The matrices $B_0(n; D, 1)$ have been introduced by Brandt in [1]. The generalized matrices $B_1(n; D, H)$ have been suggested in [5]. The traces of these are given in [5], (50) but without proof. The Brandt matrices with a character are considered here for the first time.

The arithmetical basis for extending our results to quaternion orders without restriction on H can be found in a forthcoming paper by Hijikata [7].

[1] H. Brandt, Zur Zahlentheorie der Quaternionen, Jahres-Ber. Deutsche Math.-Verein. 43 (1953), p. 23 - 57 (chap. III).

[2] C. Chevalley, Sur certains idéaux dans un algèbre simple, Abh. Math. Sem. Hamburger Univ. 10 (1934), p. 83 - 105.

[3] M. Deuring, Algebren, Ergebn. d. Math. IV, 1, Springer Verlag Berlin 1935, chapter 6.

[4] M. Eichler, Zur Zahlentheorie der Quaternionen-Algebren, Journ. reine angew. Math. 195 (1956), p. 127 - 151.

[5] M. Eichler, Quadratische Formen und Modulfunktionen, Acta Arithmetica 4 (1958), p. 23 - 57.

[6] H. Hasse, Über gewisse Ideale in einer einfachen Algebra, Act. Sci. Ind. Paris 109 (1934), p. 12 - 16.

[7] H. Hijikata, Explicite formula of the trace of Hecke operators for $\Gamma_0(N)$, to appear.

[8] M. Noether, Zerfallende verschränkte Produkte und ihre Maximalordnungen, Act. Sci. Ind. Paris 148 (1934).

Eich-42

Chapter III: THE TRACES OF THE HECKE OPERATORS

§1. Summary of the theory of generalized Abelian integrals

In this section Γ means (exceptionally) a group of linear transformations

$$z \to G(z) = \frac{az + b}{cz + d}, \qquad |G| = ad - bc = 1.$$

of the complex upper half plane H with the following properties:

a) Γ has a fundamental region F whose points correspond 1 - 1 with those of the quotient $\Gamma \setminus H$. The hyperbolic area is finite.

b) F has at most finitely many cusps. These are attached to classes of parabolic elements of Γ. A parabolic element is of the form

$$G = M^{-1}\begin{pmatrix} 1 & \lambda \\ 0 & 1 \end{pmatrix} M \tag{1}$$

with a real λ and an $M \in SL(2, R)$.

c) As a consequence of a) and b), Γ has a finite system of generators which can even be assumed in a special form.

The cusps are usually added to H which yields the amplified upper half plane $\overline{H}$ (which now depends on Γ). Γ has in $\overline{H}$ a fundamental region $\overline{F}$ which is $\Gamma \setminus \overline{H}$. The latter can be made a finite Riemann surface by a suitable definition of the local uniformizing variables in the cusps and other exceptional points.

A meromorphic function in H is said to be an automorphic form with respect to Γ of weight k if it satisfies the functional equations

$$f(z)[G]^{-k} = f(G(z))\,(cz+d)^{-k} = f(z), \qquad G \in \Gamma \tag{2}$$

the meromorphic condition must include the cusps, i.e. for a parabolic element G of the form (1) it is required that

$$f(z)[M]^{-k} = \sum_{n=n_0}^{\infty} c_n\, e^{2\pi i|\lambda|^{-1} n z} \tag{3}$$

with some negative, zero, or positive n_0.

We shall call f(z) an automorphic form of the 1st kind if f(z) is holomorphic in H and has expansions (3) with $n_0 > 0$ in all cusps. Such forms are also called cusp forms. We shall call f(z) an automorphic form of 2nd kind, if the meromorphic parts in all poles $z = z_i \in H$ are

$$f(z) = \frac{c_0}{(z-z_i)^k} + \frac{c_1}{(z-z_i)^{k+1}} + \dots + \frac{c_{n_i}}{(z-z_i)^{k+n_i}}, \tag{4}$$

while on the behaviour in the cusps nothing more is supposed than (3). All other automorphic forms are of 3rd kind. We will not deal with them.

The automorphic forms of weight 0 are the automorphic functions. They form a finite algebraic function field with C as constant field and $\Gamma \setminus \bar{H}$ as the Riemann surface. The functional equations (2) of automorphic forms are identical with

$$f(G(z)) \left(d(G(z))\right)^{k/2} = f(z) \left(dz\right)^{k/2}.$$

Thus the automorphic forms of weight k represent differentials of degree $\frac{k}{2}$. This fact allows to attach to them divisors of the function field. The theorem of Riemann-Roch then leads to the determination of the number of linearly independent automorphic forms of 1st kind (s. [11], chapt. 2) if $k \geq 2$. For k = 1 the theorem of Riemann-Roch becomes a tautology, and we have only very limited knowledge on this number (see also chapter V).

To an automorphic form of 1st or 2nd kind we attach the indefinite integral

$$F(z) = \frac{1}{(k-2)!} \int_{z_0}^{z} f(\zeta)\,(z-\zeta)^{k-2}\,d\zeta + C(z) \tag{5}$$

where C(z) is an arbitrary polynomial of degree k-2. A change of the lower limit z_0 amounts to a change of C(z). Evidently

$$\frac{d^{k-1}}{dz^{k-1}} F(z) = f(z),$$

therefore C(z) may be considered as the "integration constant".

The assumption on the automorphic forms of 1st and 2nd kinds

Eich-44

implies that the integrals (5) are one-valued functions in $\mathcal{H}$.

Applying an element $G = \begin{pmatrix} a & b \\ c & d \end{pmatrix} \in \Gamma$ in the way

$$F(z)[G]^{k-2} = F(G(z))\,(cz+d)^{k-2} \tag{6}$$

one finds

$$f(z)[G]^{k-2} = \frac{1}{(k-2)!}\int_{z_0}^{G(z)} f(\zeta)(G(z)-\zeta)^{k-2}(cz+d)^{k-2}d\zeta + C(z)[G]^{k-2}$$

$$= \frac{1}{(k-2)!}\int_{G^{-1}(z_0)}^{z} \left(f(\zeta')[G]^{-k}\right)\left((G(z)-G(\zeta'))(cz+d)(c\zeta'+d)\right)^{k-2}d\zeta'$$

$$+ C(z)[G]^{k-2}$$

and because $|G| = 1$:

$$(G(z) - G(\zeta'))(cz+d)(c\zeta'+d) = z - \zeta'.$$

Thus we arrive at

$$F(z)[G]^{k-2} = F(z) + P_G(z) \tag{7}$$

with

$$P_G(z) = \frac{1}{(k-2)!}\int_{G^{-1}(z_0)}^{z_0} f(\zeta)\,(z-\zeta)^{k-2}d\zeta + C(z)[G]^{k-2} - C(z). \tag{8}$$

(7) exhibits $F(z)$ as a periodic function with respect to the group Γ and with periods $P_G(z)$ which are polynomials of degree $k-2$. Applying (7) with 2 elements $G_1, G_2 \in \Gamma$ we get immediately

$$P_{G_1G_2}(z) = P_{G_1}(z)[G_2]^{k-2} + P_{G_2}(z) \tag{9}$$

with the operator $[G]^{k-2}$ defined by (6).

A mapping $G \to P_G(z)$ is a <u>cocycle</u>. Special cocycles are the <u>coboundaries</u> $C(z)[G]^{k-2} - C(z)$. The cocycles and the coboundaries form C-modules $C^1_{k-2}(\Gamma)$ and $B^1_{k-2}(\Gamma)$. The elements of the quotient module

$$H^1_{k-2}(\Gamma) = C^1_{k-2}(\Gamma)/B^1_{k-2}(\Gamma)$$

are the <u>cohomology classes</u>.

By (7), to each automorphic form of 1st or 2nd kind a cohomology class is attached, namely a change of the integration constant in (5) does not effect the cohomology class.

If $F(z)$ is an automorphic form of weight $2-k$, i.e. if it satisfies (7) with periods 0, its $(k-1)$-fold derivative is an automorphic form $f(z)$ of 2nd kind and weight k. (For the proof one expresses $F(z)$ by Cauchy's theorem

$$F(z) = \frac{1}{2\pi i} \oint F(\zeta) \frac{d\zeta}{(\zeta - z)}$$

and differentiates under the integral sign). Such automorphic forms $f(z)$ are called exact, and their integrals exact integrals.

Proposition 1. *Let* $M_k(\Gamma)$ *be the* C*-module of all automorphic forms of 2nd kind and weight* k *and* $M_k^e(\Gamma)$ *the submodule of exact automorphic forms. On the other hand, let* $H^1_{k-2}(\Gamma)$ *be the* $\mathbf{C}$*-module of all cohomology classes. Then (7) yields a bijective mapping*

$$M_k(\Gamma)/M_k^e(\Gamma) \leftrightarrow H^1_{k-2}(\Gamma). \tag{10}$$

The automorphic forms of 1st kind are even uniquely determined by their periods.

Let $f(z)$, $g(z)$ be two automorphic forms of 1st or 2nd kinds. We define a scalar product by the sum of the residues in a fundamental domain

$$(f(z), g(z)) = \sum \text{Res}\, \big(f(z)\, G(z)\, dz\big), \tag{11}$$

where $G(z)$ is the integral (5) of $g(z)$.

Proposition 2. *The product does neither depend on the choice of the fundamental domain nor on the integration constant. It has the symmetry property*

$$(f(z), g(z)) = (-1)^{k-1} \big(g(z), f(z)\big). \tag{12}$$

(12) follows easily by $(k-1)$-fold integration by parts.

Proposition 3. *The product (11) is 0 for arbitrary* $f(z)$ *and exact* $g(z)$, *and also for pairs* $f(z)$, $g(z)$ *of automorphic forms of 1st kinds.*

If, for a given $f(z)$ *of 2nd kind,* $(f(z), g(z)) = 0$ *for all* $g(z)$ *of*

Eich-46

1st kind, $f(z) = f_1(z) + f_2(z)$ *with* $f_1(z)$ *of first kind and* $f_2(z)$ *exact.*

We shall make the following application. Let $f_1(z),\dots,f_G(z)$ be a basis of the space $S_k(\Gamma)$ of automorphic forms of 1st kind and weight k. There exist automorphic forms $g_1(z),\dots,g_G(z)$ of second kind such that

$$(f_i(z),\ g_j(z)) = \begin{cases} 1 & \text{for } i = j, \\ 0 & \text{for } i \neq j, \end{cases}$$

and the $f_i(z)$, $g_i(z)$ together form a basis of the residue module $M_k(\Gamma)/M_k^e(\Gamma)$.

§2 Greens function, $k > 2$

Although wide generalizations are possible also in the following, we only apply §1 in the case that Γ is the congruence subgroup of the modular group defined by

$$\Gamma_0(N) = \{\begin{pmatrix} a & b \\ c & d \end{pmatrix}:\ a,\ b,\ \tfrac{c}{N},\ d \in \mathbb{Z} \text{ and } ad - bc = 1\}.$$

Of course, we mean the homogeneous group, attached to the linear transformations $z \to G(z)$ of the upper half plane, and therefore $\pm\begin{pmatrix} 1 & 0 \\ 0 & 1 \end{pmatrix}$ are the same elements.

Let

$$\Gamma_{01}(N) = \{\begin{pmatrix} a & b \\ c & d \end{pmatrix} \in \Gamma_0(N):\ d \equiv \pm 1 \bmod N\}.$$

The C-modules of automorphic forms with respect to $\Gamma_{01}(N)$ split up into direct sums

$$S_k(\Gamma_{01}(N)) = \oplus\ S_k(\Gamma_0(N),\chi), \quad M_k(\Gamma_{01}(N)) = \oplus\ M_k(\Gamma_0(N),\chi),$$

attached to the characters

$$\chi(G) = \chi\begin{pmatrix} a & b \\ c & d \end{pmatrix} = \chi(a)$$

of $(\mathbb{Z}/N\mathbb{Z})^*$, whose functions behave under $\Gamma_0(N)$ as follows:

$$f(z)[G]^{-k} = \chi(G)\ f(z), \qquad G \in \Gamma_0(N). \tag{13}$$

(This is even known in a more general situation).

We agree to use the character for all $a \in \mathbb{Z}$, namely

$$\chi(a) = 0 \quad \text{for} \quad (a, N) > 1,$$

even if $\chi(a) = 1$ for all d with $(a, N) = 1$. This convention is important in the theory of Hecke operators, and it has also been observed in connection with the Brandt matrices.

Now (7) and (9) are modified to

$$F(z)[G]^{k-2} = \chi(G)\, F(z) + P_G(z) \tag{14}$$

and

$$P_{G_1G_2}(z) = P_{G_1}(z)\, G_2^{k-2} + \chi(G_1)\, P_{G_2}(z). \tag{15}$$

The scalar products (11) are defined between a modular form f(z) of a character χ and a modular form g(z) of the complex conjugate character $\overline{\chi}$. Namely for an $H \in \Gamma_0(N)$ we have

$$\operatorname{Res}\big(f(z)\, G(z)\, dz\big) = \operatorname{Res}\big(\overline{\chi}(H) f(H(z))\big(\chi(H) G(H(z)) + P_H(z)\big)\, dH(z)\big)$$

With this proviso propositions 1 - 3 are valid for modular forms with respect to $\Gamma_0(N)$ and a character.

Greens function $K(z,\zeta)$ depends on two variables z and ζ, and is defined by the properties:

a) $K(z,\zeta)$ is a modular form of 2nd kind in z, of weight k and character χ.

b) $K(z,\zeta)$ has poles of order 1 in the points $z = G(\zeta)$ for all $G \in \Gamma_0(N)$ with the meromorphic parts

$$\frac{\chi(G)}{z-G(\zeta)}\,(c_G\zeta + d_G)^{k-2}.$$

In all other points $K(z,\zeta)$ is holomorphic. In the cusps it behaves like a cusp form, for instance $K(i\infty,\zeta) = 0$.

c)
$$\frac{\partial^{k-1}}{\partial\zeta^{k-1}} K(z,\zeta,\chi) = (-1)^k \frac{\partial^{k-1}}{\partial z^{k-1}} K(\zeta,z,\overline{\chi}).$$

Conditions a) and b) are compatible because, for an $H \in \Gamma_0(N)$,

Eich-48

$$\frac{\chi(H^{-1})\chi(G)}{H(z)-G(\zeta)}\,\frac{(c_G\zeta+d_G)^{k-2}}{(c_Hz+d_H)^k} = \frac{\chi(H^{-1}G)}{z-H^{-1}G(\zeta)}\left(\frac{c_G\zeta+d_G}{c_HH^{-1}G(\zeta)+d_H}\right)^{k-2} + \ldots$$

(+ a holomorphic function in $z = H^{-1}G(\zeta)$).

$K(z,\zeta)$ is explicitely given by the Poincaré series

$$K(z,\zeta) = \sum \frac{\overline{\chi}(G)}{G(z)-\zeta}\,(c_Gz+d_G)^{-k}, \tag{16}$$

summed over all $G \in \Gamma_0(n)$. It has to be noted that $\pm G$ yield the same summand, because $\chi(-1) = (-1)^k$. For the uniform convergence for z in a compact set, we have to arrange the sum in a suitable way. All G with the same second row are

$$G = \begin{pmatrix} a & b \\ c & d \end{pmatrix} = \begin{pmatrix} 1 & m \\ 0 & 1 \end{pmatrix}\begin{pmatrix} a_0 & b_0 \\ c & d \end{pmatrix} = \begin{pmatrix} 1 & m \\ 0 & 1 \end{pmatrix} G_0.$$

Now we demand

$$K(z,\zeta) = \sum\nolimits_{c,d}\left(\frac{\overline{\chi}(a_0)}{G_0(z)-\zeta} + \sum\nolimits_{m=1}^{\infty}\left(\frac{\overline{\chi}(a_0)}{G_0(z)+m-\zeta} + \frac{\overline{\chi}(a_0)}{G_0(z)-m-\zeta}\right)\right)(cz+d)^{-k},$$

and the uniform convergence of this series is well known.

Application of an H operates on G and G_0 without changing the factor $\begin{pmatrix} 1 & m \\ 0 & 1 \end{pmatrix}$. Therefore $K(z,\zeta)$ is a modular form of weight k and character χ.

For $z \to i\infty$, $K(z,\zeta)$ becomes 0. Applying an $M \in SL(2, Q)$ and afterward letting $z \to i\infty$ shows that $K(z,\zeta)$ vanishes in all cusps.

In $z = G^{-1}(\zeta)$ we have the wanted singularities which is checked by the same consideration as the compatibility of a) and b).

Eventually we find by (k-1)-fold differentiation

$$\frac{\partial^{k-1}}{\partial\zeta^{k-1}}K(z,\zeta) = \sum \frac{(k-1)!\,\overline{\chi}(G)}{\left((G(z)-\zeta)(c_Gz+d_G)\right)^k} = \sum \frac{(k-1)!\;\chi(G^{-1})}{\left((z-G^{-1}(\zeta))(c_{G^{-1}}\zeta+d_{G^{-1}})\right)^k},$$

in accordance with the property c).

<u>Proposition 4.</u> *For* $k > 2$*, Green's function has the following property (besides a) - c)):*

$$K(z,\zeta) = -\sum f_i(z)\, G_i(\zeta) + L(z,\zeta)$$

where $f_i(z)$ *is a basis of the* C-*module* $S_k(\Gamma_0(N),\chi)$ *of modular forms of weight* k, *character* χ, *and of 1st kind; the* $G_i(\zeta)$ *are integrals of modular forms of weight* k, *character* $\overline{\chi}$, *and of 2nd kind, and with the scalar products*

$$(f_i(z),\, g_j(z)) = \begin{cases} 1 & \text{for } i = j, \\ 0 & \text{for } i \neq j, \end{cases}$$

and $L(z,\zeta)$ *is a modular form in* z *of weight* k *and character* χ, *and a modular form of weight* 2-k *and character* $\overline{\chi}$ *in* ζ *(i.e. an exact integral).*

The proof rests on the following lemma which will be commented on in §8.

Lemma. Let $G = \begin{pmatrix} a & b \\ c & d \end{pmatrix} \to P_G(z)$ be a cocycle of $\Gamma_0(N)$ in the polynomials of degree k-2. For a compact set $K \subset H$ there exists a constant $C(K)$ such that for all $z \in K$ and all G holds

$$|P_G(z)| < C(K)\,(c^2 + d^2)^{\frac{k}{2}-1} \log(c^2 + d^2).$$

Property c) of Green's function shows that it is an integral of 2nd kind in the variable ζ, with character $\overline{\chi}$. Therefore

$$K(z,\zeta)[G]^{k-2} = \overline{\chi}(G)\, K(z,\zeta) + P_G(\zeta;\, z),$$

and the periods are polynomials in ζ of degree k-2 with coefficients which are modular forms of 1st kind in z. As functions in ζ they satisfy the cocycle condition (15) with $\overline{\chi}$ instead of χ. With the basis $f_i(z)$ of $S_k(\Gamma_0(N),\chi)$ we can write

$$P_G(\zeta;\, z) = \sum f_i(z)\, P_{i,G}(\zeta),$$

and the $P_{i,G}(\zeta)$ are also cocycles. We want to show that there exist integrals $G_i(\zeta)$ of 2nd kind with periods $-P_{i,G}(\zeta)$.

Eich-50

If such $G_i(\zeta)$ do exist, $K(z,\zeta) + \sum f_i(z)\, G_i(\zeta) = L(z,\zeta)$ is an exact integral in ζ and of course a modular form in z. Because of the property b) of $K(z,\zeta)$ we have for each modular form $f_i(z)$ of 1st kind

$$\sum \mathrm{Res}\big(K(z,\zeta)\, f_i(\zeta)\, d\zeta\big) = \mathrm{Res}\Big(\frac{f_i(\zeta)\, d\zeta}{z-\zeta}\Big) = -f_i(z)$$

which implies that the $f_i(z)$, $G_j(z)$ are orthogonal.

For the construction of the $G_i(z)$ we note at first that

$$P_T(\zeta,z) = 0, \qquad T = \begin{pmatrix} 1 & 1 \\ 0 & 1 \end{pmatrix}.$$

Namely $K(z,\zeta) \to 0$ as $\zeta \to i\infty$ which follows immediately from (16). The same holds then for the $P_{i,T}(\zeta)$. Because of (15) this leads to

$$P_{i,\,T^mH}(z) = P_{i,\,H}(z) \qquad (m = 0, \pm 1, \ldots)$$

Now the $G_i(z)$ are constructed by means of the generalized Poincaré series (we may drop the subscript i in the following)

$$\Phi(z) = \sum_{\substack{(c,d)=1 \\ c\equiv 0 \bmod N}} \frac{\chi(d) P_H(z)}{(cz+d)^{k+2}} = \sum_H \frac{\chi(H)\; P_H(z)}{(c_H z+d_H)^{k+2}},$$

where $H = \begin{pmatrix} a & b \\ c & d \end{pmatrix}$ is an element of $\Gamma_0(N)$ with 2nd row c, d. The sum converges uniformly for all z in a compact set by virtue of the lemma. The cocycle condition (14) implies

$$\Phi(z)[G]^{-4} = \sum_H \frac{\chi(H)\; P_H(z)\; [G]^{k-2}}{(c_{HG}z+d_{HG})^{k+2}}$$

$$= \bar{\chi}(G) \sum_{HG} \frac{\chi(HG)\; P_{HG}(z)}{(c_{HG}z+d_{HG})^{k+2}} + P_G(z) \sum_{HG} \frac{1}{(c_{HG}z+d_{HG})^{k+2}}.$$

Therefore, with

$$\phi(z) = \sum_H \frac{1}{(c_H z+d_H)^{k+2}}, \qquad G(z) = -\frac{\Phi(z)}{\phi(z)}$$

has the desired property

$$G(z)[G]^{k-2} = \bar{\chi}(G)\; G(z) - P_G(z).$$

§3 Green's function, $k = 2$

A function $K(z,\zeta)$ with all 3 properties a) - c) does not exist in general if $k < 2$. Now we can only demand

a) $K(z,\zeta)$ is a modular form of 2nd kind in z, of weight 2 and character χ.

b) $K(z,\zeta)$ has poles of order 1 in the points $z = G(\zeta)$ for all $G \in \Gamma_0(N)$ with the meromorphic parts

$$\frac{\chi(G)}{z - G(\zeta)}.$$

If $\chi(G)$ is not the principal character $\chi(G) = 1$, $K(z,\zeta)$ is holomorphic in all other points of H and vanishes in all cusps. If $\chi(G)$ is the principal character, $K(z,\zeta)$ has one further pole in the set $z = G(z_0)$ with an arbitrary z_0 with residue -1. Otherwise it has the same properties.

We prove the existence by means of the theorem of Riemann-Roch for the field of modular functions with respect to the group $\Gamma_0(N,\chi) = \{G \in \Gamma_0(N): \chi(G) = 1\}$. There exists a differential $K_1(z,\zeta)\,dz$, which has poles of orders and residues 1 and -1 in $z = \zeta$ and another (arbitrary) point z_0. If χ is the principal character, $K_1(z,\zeta)$ has the wanted properties. If χ is not the principal character, let G_ν be a system of representatives of the cosets of $\Gamma_0(N,\chi)$ in $\Gamma_0(N)$. Now we have $\sum \chi(G_\nu) = 0$. With these G_ν we put

$$K_2(z,\zeta)\,dz = \sum_\nu \overline{\chi}\,(G_\nu)\,K_1(G_\nu(z),\zeta)\,dG_\nu(z)$$

which evidently behaves in the wanted way in the poles $z = G(\zeta)$. It has further poles with residues $-\chi(G_\nu)$ in the points $z = G_\nu(z_0)$. These are not equivalent with respect to $\Gamma_0(N,\chi)$, and the sum of residues is $\sum \chi(G_\nu) = 0$. Therefore there exists a differential $K_3(z)\,dz$ with these poles and residues, and

$$K(z,\zeta) = K_2(z,\zeta) + K_3(z)$$

Eich-52

has only poles at $z = G(\zeta)$.

We cannot prove that $K(z,\zeta)$ is an integral of a modular form in ζ. But from b) follows that

$$K(z,\zeta)[G]_\zeta^0 = \overline{\chi}(G)\ K(z,\zeta) + P_G(\zeta;\ z)$$

where $P_G(\zeta;\ z)$ is a holomorphic modular form in z of weight 2, vanishing in the cusps (because $K(z,\zeta)\ dz$ is integral in the cusps). As function in ζ, $K(z,\zeta)$ is limited for each z, and therefore the $P_G(\zeta;\ z)$ are also limited. Eventually

$$K(z,\zeta + 1) = K(z,\zeta) \quad \text{whence } P_T(\zeta;\ z) = 0.$$

Now we have

$$P_G(\zeta;\ z) = \sum f_i(z)\ P_{i,G}(\zeta)$$

with analogue functions $P_{i,G}(\zeta)$. As in No. 2, we can construct meromorphic functions $G_i(\zeta)$ with the periods

$$G_i(\zeta)[G]^0 - \overline{\chi}(G)\ G_i(\zeta) = P_{i,\ G}(\zeta),$$

and thus we arrive at

Proposition 5. *For* $k = 2$, *Greens function has the following properties (besides a) and b)):*

$$K(z,\zeta) = -\sum f_i(z)\ G_i(\zeta) + L(z,\zeta),$$

where $L(z,\zeta)$ *is a modular form of weight* 2 *and character* χ *in* z, *and of weight* 0 *and character* $\overline{\chi}$ *in* ζ, *and where the* $G_j(z)$ *are meromorphic functions with*

$$\sum \mathit{Res}(f_i(z)\ G_j(z)\ dz) = \begin{cases} 1 & \text{for } i = j, \\ 0 & \text{for } i \neq j. \end{cases}$$

The last line is verified as in §2.

§4 The action of the correspondences on Greens function

The modular correspondences or Hecke operators T(n) on modular forms f(z) of weight k and character χ are defined by

$$f(z)\, dz^{\frac{k}{2}}\, T(n) = n^{\frac{k}{2}-1} \sum \bar{\chi}(a)\, f(\frac{az+b}{d})\, (d\frac{az+b}{d})^{\frac{k}{2}}$$

or

$$(17)\quad f(z)[T(n)]^{-k} = n^{\frac{k}{2}-1} \sum \bar{\chi}(a)\, f(\frac{az+b}{d}) (\frac{a}{d})^{\frac{k}{2}} = \sum \frac{1}{d}\bar{\chi}(a)\, f(\frac{az+b}{d})\, a^{k-1}$$

where a, b, d, run over the following system of integers:

$$ad = n, \quad a > 0, \text{ and } b \bmod d.$$

The last expression shows that we may replace a, b, d by -a, -b, -d without changing the result. The same will be observed in the following. If n is a square, we split up the operator into the sum

$$(18)\qquad T(n) = T(n)' + (\sqrt{n})^{k-2}\, \chi(\sqrt{n}) \times I \qquad (\sqrt{n} > 0)$$

where $T(n)'$ means the same as (17), but without the term $a = d$, $b = 0$, and I the unit correspondence.

We apply $T(n)'$ to Greens function with respect to the variable z and afterwards put $\zeta = z$:

$$[K(z,\zeta)[T(n)']_z^{-k}]_{\zeta=z} = -\sum (f_i(z)[T(n)']^{-k})\, G_i(z) + h_n(z),$$

and we derive from propositions 4 and 5 that $h_n(z)$ is a modular form of weight 2. On the right we have

$$\sum f_i(z)[T(n)']^{-k} = \sum_j t_{ij}\, f_j(z)$$

with the representation (t_{ij}) of $T(n)'$ in the **C**-module $S_k(\Gamma_0(N),\chi)$. The trace of this matrix we shall denote by $s_k(T(n)',\chi)$; we can express it by virtue of propositions 4 and 5 as:

$$\sum \operatorname{Res}\Big(\sum_i \big(f_i(z)[T(n)']^{-k}\big) G_i(z)\Big)\, dz = s_k(T(n)',\chi).$$

Thus we have the sum of the residues of poles in a fundamental region

$$(19)\qquad \sum \operatorname{Res}[\operatorname{Res} K(z,\zeta)[T(n)']_z^{-k}]_{\zeta=z}\, dz = -s_k(T(n)',\chi).$$

We now evaluate the left hand side. Because of property b) of

Eich-54

$K(z,\zeta)$ it is

$$\sum \text{Res } \frac{\overline{\chi}(a)\, a^{k-1}}{d} \frac{\chi(G)\, dz}{\frac{az+b}{d} - G(z)} - (\textstyle\sum' t)$$

where the second summand (...) occurs only if $k = 2$ and $\chi(G)$ is the principal character; and then it is equal to the number of summands in (17) with $T(n)'$ instead of $T(n)$, namely the sum of all positive divisors t of n, with $(N, nt^{-1}) = 1$, except that for $n = t^2$, t must be replaced by $t-1$.

Another description of the Hecke operator $T(n)$ is ([2], chap. V and [11], 3.4)

$$T(n)' = \sum_{t^2/n} t^{k-2}\, \overline{\chi}(t)\, T_0(nt^{-2}), \qquad (t^2 \neq n)$$

where $T_0(n)$ is defined similarly, only with relatively prime a, b, d. The correspondences $T_0(n)$ are the double cosets

$$T_0(n) = \Gamma_0(N)\, M_0(n)\, \Gamma_0(N), \qquad M_0(n) = \begin{pmatrix} 1 & 0 \\ 0 & n \end{pmatrix}$$

They can also be written

$$T_0(n) = \cup_\nu \Gamma_0(N)\, M_0(n)\, G_\nu$$

where G_ν is a system of representatives of the right cosets in

$$\Gamma_0(N) = \cup_\nu \left(\Gamma_0(N) \cap M_0(n)^{-1} \Gamma_0(N)\, M_0(n)\right) G_\nu .$$

These cosets coincide with the $\Gamma_0(N) \begin{pmatrix} a & b \\ 0 & d \end{pmatrix}$, where $(a, N) = 1$. Thus we have

$$f(z)[T_0(n)]^{-k} = n^{\frac{k}{2}-1} \sum_\nu \overline{\chi}(G_\nu)\, f(M_0(n)\, G_\nu(z)),$$

because

$$\overline{\chi}(G_\nu) = \overline{\chi}(A) \quad \text{with A as in } M_0(n)\, G_\nu = \begin{pmatrix} A & B \\ C & D \end{pmatrix}.$$

Now we can express (19) as follows:

<u>Proposition 6.</u> *The trace of the Hecke operator* $T(n)'$ *in* $S_k(\Gamma_0(N), \chi)$ *is*

$$s_k(T(n)', \chi) = -n^{\frac{k}{2}-1} \sum \text{Res } \frac{\overline{\chi}(A)\, dz}{\frac{Az+B}{Cz+D} - z} \left(\frac{\sqrt{n}}{Cz+D}\right)^k + (\textstyle\sum' t)$$

Where $\binom{A\ B}{C\ D}$ *runs over all integral matrices with determinant* n, *which are different from* f times a *unimodular matrix, and all poles in one fundamental region have to be taken. The second term occurs only for* $k = 2$ *and the principal character* $\chi^{(G)} = 1$, and $\sum{}'_t$ *means the sum over all positive divisors* t of n *with* $(N, nt^{-1}) = 1$, *except that for* $n = t^2$, *the sum and* t *must be replaced by* t-1.

Furthermore the trace of T(n) *is*

$$s_k(T(n),\chi) = s_k(T(n)',\chi) + \begin{cases} n^{\frac{k}{2}-1}\,\overline{\chi}(\sqrt{n})\ \dim S_k(\Gamma_0(N),\chi), \\ 0, \end{cases}$$

according as n *is a rational square or not.*

§5 <u>The contribution of the cusps to the trace formula</u>

The poles occurring in proposition 6 are the <u>fixed points</u> of T(n)', i.e. the solutions of $\frac{Az+B}{Cz+D} = z$. We must treat the fixed points lying in the cusps and those in H separately.

At first we consider the cusp $i\infty$. Then $C = 0$. In order to determine the residue, we introduce the local uniformizer $q = e^{2\pi iz}$. If n is not a square, the contribution of the operator $\sum_B \left[\begin{smallmatrix} A & B \\ 0 & D\end{smallmatrix}\right]^{-k}$ to the trace of T(n)' is

$$-2\pi i\overline{\chi}(A)\,\frac{A^{k-1}}{D}\,\sum_{B \bmod D} \mathrm{Res}\left(\frac{dz}{1-e^{2\pi i(z-D^{-1}(Az+B))}}\right)$$

$$= -\overline{\chi}(|A|)|A|^{k-1}\,\mathrm{Res}\left(\frac{q^{|A|}}{q^{|A|}-q^{|D|}}\,\frac{dq}{q}\right) = -\overline{\chi}(|A|)|A|^{k-1}\begin{cases}1 & \text{for } |A| < |D|, \\ 0 & \text{for } |A| > |D|.\end{cases}$$

If n is a square, we must omit the summand $A = D$, $B = 0$. The contribution of $\sum_B \left[\begin{smallmatrix} A & B \\ 0 & A\end{smallmatrix}\right]^{-k}$ is now

$$-\overline{\chi}(\sqrt{n})\,n^{\frac{k}{2}-1}\,\frac{\sqrt{n}-1}{2} \qquad (\sqrt{n} > 0).$$

So we find the total contribution of T(n)' and the cusp i to the trace of T(n)':

$$-\sum_{\substack{t/n \\ 0<t<\sqrt{n}}} \overline{\chi}(t)t^{k-1} - \left(\overline{\chi}(\sqrt{n})\ n^{\frac{k}{2}-1}\ \frac{\sqrt{n}-1}{2}\right).$$

The second summand must be cancelled if n is not a rational square.

Eich-56

$\Gamma_0(N)\backslash H$ has 2^r cusps, where r is the number of prime factors of N, if N is square-free. They are obtained from $i\infty$ by substitutions $\begin{pmatrix}\alpha & \beta\\ \gamma & \delta\end{pmatrix} \in \Gamma$, defined by

$$(20) \qquad \begin{pmatrix}\alpha & \beta\\ \gamma & \delta\end{pmatrix} \equiv \begin{cases}\begin{pmatrix}1 & 0\\ 0 & 1\end{pmatrix} \bmod N_1,\\ \begin{pmatrix}0 & 1\\ -1 & 0\end{pmatrix} \bmod \frac{N}{N_1}\end{cases}$$

for all 2^r divisors N_1 of N.

If $\begin{pmatrix}A & B\\ C & D\end{pmatrix}$ leaves a cusp fixed which is attached to $\begin{pmatrix}\alpha & \beta\\ \gamma & \delta\end{pmatrix}$,

$$\begin{pmatrix}\alpha & \beta\\ \gamma & \delta\end{pmatrix}\begin{pmatrix}A & B\\ C & D\end{pmatrix}\begin{pmatrix}\alpha & \beta\\ \gamma & \delta\end{pmatrix}^{-1} = \begin{pmatrix}t & s\\ o & r\end{pmatrix}$$

leaves $z = i\infty$ fixed, and

$$\begin{pmatrix}A & B\\ C & D\end{pmatrix} = \begin{pmatrix}\alpha\delta t - \beta\gamma r + \gamma\delta s & *\\ * & *\end{pmatrix}.$$

We can now express $\bar{\chi}(A)$ in the following way: put

$$(21) \qquad v(t, n, N_1) \equiv \begin{cases} t \bmod N_1,\\ \frac{n}{t} \bmod \frac{N}{N_1},\end{cases}$$

then comparison of (20) and (21) yields

$$\bar{\chi}(A) = \bar{\chi}(v(t, n, N_1)).$$

So the contribution of all cusps to the trace of T(n)' is

$$-\sum_{\substack{t/n\\ 0<t<\sqrt{n}}} t^{k-1} \sum_{N_1/N} \bar{\chi}(v(t, n, N_1)) - (2^{r-1}\bar{\chi}(\sqrt{n})\sqrt{n}^{k-1}(\sqrt{n} - 1))$$

The second summand occurs only if n is a square.

§6 <u>The contribution of the fixed points in H to the trace formula</u>

A fixed point of T(n)' is a solution of the quadratic equation $M(z) = \frac{Az + B}{Cz + D} = z$ with $M = \begin{vmatrix}A & B\\ C & D\end{vmatrix} = n$; this is $z = z_0 = \frac{A-D + \sqrt{(A+D)^2-4n}}{2C}$. The contribution of it to $s_k(T(n)',\chi)$ is by proposition 6 equal to

$$\frac{1}{n}\,\frac{\bar{\chi}(A)}{1-\frac{n}{\rho^2}}\left(\frac{n}{\rho}\right)^k \operatorname{Res}\left(\frac{dz}{z-z_0}\right) = \frac{\bar{\chi}(A)}{\rho-\bar{\rho}}\,\bar{\rho}^{k-1},$$

where

(22) $$\rho = \frac{s + \sqrt{s^2 - 4n}}{2}, \qquad s = s(M) = A+D,$$

and $\bar{\rho}$ is the complex conjugate.

In evaluating the residue we have assumed that $z-z_0$ is a local uniformizer. This is indeed true except when z_0 is the fixed point of an element of finite order of $\Gamma_0(N)$. Then $(z-z_0)^2$ or $(z-z_0)^3$ are locally uniformizing variables and

$$\mathrm{Res}\,\frac{dz}{z-z_0} = \tfrac{1}{2}\mathrm{Res}\,\frac{d(z-z_0)^2}{(z-z_0)^2} = \frac{1}{2} \text{ or } = \tfrac{1}{3}\mathrm{Res}\,\frac{d(z-z_0)^3}{(z-z_0)^3} = \frac{1}{3}.$$

So we have to attach the factor $\frac{1}{2}$ or $\frac{1}{3}$ in these cases.

Now we ask, how many non-equivalent z_0 occur which are solutions of $M(z_0) = z_0$ for an M with $s(M) = s$ and $M = n$. As one easily checks, all matrices M in a field $K = Q(M)$ have the same fixed point. Therefore these fixed points correspond bijectively to the orders

$$\nu_0 = I \cap K,$$

where $I = \{\begin{pmatrix} A & B \\ C & D \end{pmatrix}: A,\ B,\ \frac{C}{N},\ D \in \mathbb{Z}\}$ and K is an imaginary quadratic field generated by a two-rowed matrix; the order ν_0 must contain an element M of trace s and determinant n. This puts us into the situation of II, §4 . There we saw (proposition 4) that the number of classes of such elements which are equivalent with respect to the unit group of I, is equal to

$$\Pi_{p|N}\,(1 + \{\frac{\Delta_0}{p}\})\ h(\Delta_0),$$

where Δ_0 is the discriminant of ν_0. We want to show that the number of classes of equivalent (with respect to $\Gamma_0(N)$) orders ν_0 is the same. $\Gamma_0(N)$ has in the unit group of I the index 2. Therefore the number of classes of elements which are equivalent with respect to $\Gamma_0(N)$ is twice as large. On the other hand, each order ν_0 contains 2 such elements, namely M and $M' = sE - M$. These are not equivalent with respect to $\Gamma_0(N)$. Indeed, if $M' = U^{-1}M\,U$ with an $U \neq E \in \Gamma_0(N)$, U^2 would

Eich-58

commute with M. And since U satisfies a quadratic equation, $U^2 = \pm E$. If $U^2 = E$, the determinant of U would be -1. If $U^2 = -E$, the matrix algebra would be generated by two elements $\Omega = M - \frac{s}{2} E$ and U with $\Omega^2 = \frac{1}{4}\Delta < 0$, $U^2 = -1$, and $\Omega U + U\Omega = 0$. This is impossible.

Proposition 4 in II gathers the classes of such orders in families. Two orders ν_0, ν_1 are connected in the way

$$\nu_1 = U_p P_p^{\mu_p} \nu_0 (U_p P_p^{\mu_p})^{-1}, \quad P_p = \begin{pmatrix} 0 & 1 \\ p & 0 \end{pmatrix}, \quad \mu_p = 0 \text{ or } 1$$

for all $p|N$ with units U_p in I_p. They are in the same family if and only if all $\mu_p = 0$ for $\{\frac{\Delta_0}{p}\} \neq 0$. As in II, §8, the families of such orders are attached to the solutions A_p, D_p of the congruences

$$(23) \qquad x^2 - sx + n \equiv (x - A_p)(x - D_p) \mod p$$

for all $p|N$. To each set $\{A_p\}$ of solutions there corresponds one family. The sum over all $\overline{\chi}(A)$ belonging to elements M of trace s and determinant n which are contained in orders ν_0 with discriminant $\Delta_0 = (s^2 - 4n) f^{-2}$ is

$$\Pi_{p|N} \tfrac{1}{2}(1+\{\frac{s^2-4n)f^{-2}}{p}\})(\overline{\chi}_p(A_p)+\overline{\chi}_p(D_p)) \frac{2h((s^2-4n)f^{-2})}{w((s^2-4n)f^{-2})}$$

where

$$(24) \qquad \chi(A) = \Pi_{p|N}\, \chi_p(A) \quad \text{and} \quad \chi_p(u) = \chi(u_p) \text{ for } u_p \equiv \begin{cases} u \mod p, \\ 1 \mod \frac{N}{p}. \end{cases}$$

Lastly we symmetrize the final sum. With an $s = s(M)$, also $-s$ occurs. Now $\frac{1}{\rho-\overline{\rho}}\, \overline{\rho}^{k-1}$ goes into $(-1)^{k-1} \frac{1}{\rho-\overline{\rho}}\, \rho^{k-1}$ with $s \to -s$. But because $\chi(-1) = (-1)^k$, also $\overline{\chi}(A)$ and the $\overline{\chi}_p(A_p)$, $\overline{\chi}_p(D_p)$ change the sign by $(-1)^k$. Therefore we can replace $\frac{1}{\rho-\overline{\rho}}\, \overline{\rho}^{k-1}$ by $\frac{-1}{2}\, \frac{\rho^{k-1}-\overline{\rho}^{k-1}}{\rho-\overline{\rho}}$ in the sum.

Taking all these remarks together we get

<u>Proposition 7.</u> *The contribution of the fixed points in H to the trace of* T(n)' *is*

Eich-59

$$-\sum_{s,f}\left(\prod_{p|N}\tfrac{1}{2}\left(1+\left\{\frac{(s^2-4n)f^{-2}}{p}\right\}\right)(\bar{\chi}_p(A_p)+\bar{\chi}_p(D_p))\right)\frac{h((s^2-4n)f^{-2})}{w((s^2-4n)f^{-2})}\,\frac{\rho^{k-1}-\bar{\rho}^{k-1}}{\rho-\bar{\rho}},$$

where s, f run over the same integers as in Ch.II, theorem 7 except $f^2 = n$ *in case that* n *is a rational square. Also* h( ) and w( ) *have the same meaning. The character* $\chi(A)$ *is decomposable into the product* (24). A_p and D_p *mean the solutions of the congruences* (23), and is defined in (22).

See p. 137 (Errata)

§7 <u>The final trace formula</u>

According to the corollary to proposition 6 we need yet the rank of $S_k(\Gamma_0(N),\chi)$. Let

(25) $\Gamma_0(N),\chi) = \{G \in \Gamma_0(N): \chi(G) = 1\}$ and $e_\chi = [\Gamma_0(N): \Gamma_0(N,\chi)]$.

The genus g of the field of modular functions with respect to $\Gamma_0(N,\chi)$ is

(26 $$g_\chi = 1 + \frac{e_\chi}{12}\prod_{p|N}(p+1) - \frac{e_\chi}{2}\prod_{p|N}2 - e_\chi\left\{\frac{1+\chi(a_2)}{8}\prod_{p|N}\left(1+\left(\frac{-4}{p}\right)\right)\right\} - e_\chi\left\{\frac{1+\chi(a_3)+\bar{\chi}(a_3)}{9}\prod_{p|N}\left(1+\left(\frac{-3}{p}\right)\right)\right\},$$

where a_2 and a_3 are solutions of the congruences

$$a_2^2 + 1 \equiv 0 \bmod N,\quad a_3^2 + a_3 + 1 \equiv 0 \bmod N.$$

The formula is independent of the solutions a_2, a_3 taken.

<u>Proof</u>. We apply the formula on p. 23 in [11]. The number of inequivalent cusps is $e_\chi \Pi 2$. The number of inequivalent elements of order 2 in $\Gamma_0(N,\chi)$ is $\Pi(1+(\frac{-4}{p}))$ or 0 according as $\chi(a_2) = 1$ or -1. Indeed such an element is $G = \begin{pmatrix} a & b \\ c & -a\end{pmatrix} \equiv \begin{pmatrix} a & b \\ 0 & -a\end{pmatrix} \bmod N$ and $\chi(G) = \chi(a)$. The number of inequivalent elements of order 3 in $\Gamma_0(N,\chi)$ is $e_\chi\Pi(1+(\frac{-3}{p}))$ or 0 according as $\chi(a_3) = 1$ or $\neq 1$. Namely such an element is $G = \begin{pmatrix} a & b \\ c & d\end{pmatrix} \equiv \begin{pmatrix} a & b \\ 0 & d\end{pmatrix} \bmod N$, $G^2 + G + E \equiv 0$ or $3E \bmod N$, and $\chi(G) = \chi(a)$.

The dimension of the space of cusp forms with respect to $\Gamma_0(N,\chi)$ is

$$\dim S_k(\Gamma_0(N,\chi)) = \sum_{\nu \bmod e_\chi} \dim S_k(\Gamma_0(N),\chi^\nu) = \sum_{t|e_\chi}\phi\left(\frac{e_\chi}{t}\right)\dim S_k(\Gamma_0(N),\chi^t),$$

since two $S_k(\Gamma_0(N),\chi^\nu)$ have the same ranks if the χ^ν have the same orders. This implies

Eich-60

(27) $$\dim S_k(\Gamma_0(N),\chi) = \phi(e_\chi)^{-1}\sum_{t\mid e_\chi} \mu(t) \times \dim S_k(\Gamma_0(N),\chi^t)),$$

where $\mu(t)$ is the Möbius function.

In the applications of the trace formula which we will make in IV, we only need the case of even k. (For odd k, one has to distinguish between regular and irregular cusps which we would like to avoid; see [11], p. 29, 47). For even k, the dimension is given by theorem 2.24, p. 46, in [11]:

(28) $$\dim S_k(\Gamma_0(N,\chi)) =$$

$$=\begin{cases} g_\chi & \\ (k-1)(g_\chi-1)+(\frac{k}{2}-1)\Pi_{p\mid N}\, 2+[\frac{k}{4}]\frac{1+\chi(a_2)}{2}\Pi_{p\mid N}\,(1+(\frac{-4}{p}))+[\frac{k}{3}]\frac{1+\chi(a_3)+\bar\chi(a_3)}{3} \times & \\ \times\ \Pi_{p\mid N}(1+(\frac{-3}{p})) & \end{cases} \begin{matrix} \\ \text{for } k = 2, \\ \text{for } k > 2. \end{matrix}$$

Especially we have

(29) $$\dim S_k(\Gamma_0(N)) = \dim S_k(\Gamma_0(N),1) =$$

$$=\begin{cases} 1+\frac{1}{12}\Pi_{p\mid N}(p+1)-\frac{1}{2}\Pi_{p\mid N}2-\frac{1}{4}\Pi_{p\mid N}(1+(\frac{-4}{p}))-\frac{1}{3}\Pi_{p\mid N}(1+(\frac{-3}{p})) & \text{for } k = 2, \\ \frac{k-1}{12}\Pi_{p\mid N}(p+1)-\frac{1}{2}\Pi_{p\mid N}2-(\frac{k-1}{4}-[\frac{k}{4}])\Pi_{p\mid N}(1+(\frac{-4}{p}))-(\frac{k-1}{3}-[\frac{k}{3}]\Pi_{p\mid N}(1+(\frac{-3}{p})) & \text{for even } k, > 2. \end{cases}$$

We collect the results of §4 - §6:

<u>Theorem.</u> *The trace of* T(n) *in the space* $S_k(\Gamma_0(N),\chi)$ *is, under the assumptions that* N *is the product of different primes and* $k > 1$

$$s_k(T(n),\chi)$$

$$=-\sum_{s,f} \frac{\rho^{k-1}-\bar\rho^{k-1}}{\rho-\bar\rho}\,\frac{h((s^2-4n)f^{-2})}{w((s^2-4n)f^{-2})}\,\Pi_{p\mid N}\frac{1}{2}(1+\{\frac{(s^2-4n)f^{-2}}{p}\})(\bar\chi_p(A_p)+\bar\chi_p(D_p))$$

$$-\sum_{t\mid n}t^{k-1}\sum_{N_1\mid N}\bar\chi(v(t,\ n,\ N_1)),$$

summed over all positive divisors t *of* n *which are* $< \sqrt{n}$, *with* $v(t, n, N_1)$ *in (21),*

$$+\begin{cases} n^{\frac{k}{2}-1}\bar\chi(\sqrt{n})\left(\dim S_k(\Gamma_0(N),\chi)-\frac{\sqrt{n}-1}{2}\Pi_{p\mid N}2\right), & \textit{for } \sqrt{n}\equiv 0 \bmod 1 \\ 0 & \textit{otherwise,} \end{cases}$$

Eich-61

$$+\begin{cases} \sum t, & \textit{summed over all positive divisors } t \textit{ of } n \textit{ with } (N, n\,t^{-1}) = 1, \\ & \textit{if } k = 2 \textit{ and } \chi \textit{ is the principal character} \\ 0 & \textit{otherwise,} \end{cases}$$

$$-\begin{cases} 1 & \textit{if } k = 2,\ \chi \textit{ is the principal character, and } n \textit{ is a rational} \\ & \textit{square and prime to } N, \\ 0 & \textit{in all other cases.} \end{cases}$$

In the first line, s *and* f *run over all integers with*

$$-2\sqrt{n} < s < 2\sqrt{n},\ (s^2 - 4n)f^{-2} \equiv 0 \text{ or } 1 \bmod 4,\ 0 < f < n$$

The character χ *is decomposed into the product (24), and* A_p, D_p *mean the solutions of the congruences (23). The dimension of* $S_k(\Gamma_0(N),\chi)$ *is determined by (26) - (29), if* k *is even.* $h(\Delta)$ *and* $w(\Delta)$ *mean the numbers of ideal classes and of units of the order of discriminant* Δ, *and* $\rho = \frac{1}{2}(s+\sqrt{s^2-4n})$, $\bar{\rho} = \frac{1}{2}(s-\sqrt{s^2-4n})$.

§8 Remarks on the literature

Our method of determining the trace of T(n) has been developed by the author [1]. Instead of the generalized abelian integrals a more abstract concept can be used for the same purpose: namely systems of principal parts or sets of differentials $w_p\, dp^{1-\frac{k}{2}}$ for all points of the Riemann surface, expressed in local uniformizers p, in which only the meromorphic parts of the w_p play a rôle (Kappus [6]). This procedure has the advantage that correspondences and their representations can be studied in algebraic function fields over constant fields with prime characteristic. It is carried out in [2] (chap. III, §5 and chap. V, §2).

The lemma in §2 has been proved in [3]. The basic idea is to partition H into equivalent fundamental regions G(F), and to join F and G(F) by a chain of neighbouring fundamental regions. This yields an expression of G as a product of generators of the group, consisting of approximately (F, G(F)) factors, where (F, G(F)) is the hyperbolic

Eich-62

distance between these regions.

The lemma has been used by Lehner [8] in constructing generalized abelian integrals under more general conditions. A quite different method for the same purpose has been given by Knopp [7]. This attaches to a given system of periods even an integral which is everywhere holomorphic with the exception of one cusp. See also Husseini and Knopp [5].

Shimura [11, chap. 8] discusses the connections between modular forms of 1st kind and their periods in yet a third way, and under much more general conditions. It is an important result that these systems of periods often allow to construct Abelian varieties, in generalization of the known connection between the ordinary abelian integrals of 1st kind and the Jacobian variety.

Hijikata [4] determines the traces of Hecke operators without the restriction to square-free level N, and also for odd k. He includes the automorphic forms attached to unit groups of indefinite quaternion algebras.

So far the analytic (or algebraic) basis of the trace formula does not allow to treat problems in higher dimensions. But such problems are approachable by Selberg's general trace formula [9]. Indeed Shimizu [10] has worked out cases of Hilbert modular forms.

[1] M. Eichler, Eine Verallgemeinerung der Ableschen Integrale, Math. Zeitschr. 67 (1957), p. 267 - 298.

[2] ---------- Einführung in die Theorie der algebraischen Zahlen und Funktionen, Birkhäuser Verlag Basel 1963. -English translation Academic Press, New York and London 1966.

[3] ---------- Grenzkreisgruppen und kettenbruchartige Algorithmen, Acta Arithmetica 11 (1965), p. 169 - 180.

[4] H. Hijikata, Explicite formula of the trace of Hecke operators, to appear in the Journ. Math. Soc. Japan.

[5] S. Y. Husseini and M. I. Knopp, Eichler cohomology and automorphic forms, Illinois Journ. Maths. 15 (1971), p. 565 - 577.

[6] H. Kappus, Darstellungen von Korrespondenzen algebraischer Funktionenkörper und ihre Spuren, Journ reine angew. Math. 210 (1962), p. 123 - 140.

[7] M. I. Knopp, Construction of automorphic forms on H-groups and supplementary Fourier series, Trans. Amer. Math. Soc. 103 (1962), p. 168 - 188.

[8] J. Lehner, Automorphic integrals with preassigned period polynomials and Eichler cohomology, Nat. Bur. of standarts (1971), p. 49 - 56.

[9] A. Selberg, Automorphic functions and integral operators, Seminars on analytic functions, Inst. for Advanced Study, Princeton, N. J. and U.S. Air Force Office of Scientific Research, vol. 2 (1957), p. 152 - 161.

[10] H. Shimizu, On traces of Hecke operators, Journ. Fac. Sci. Univ. Tokoyo 10 (1965), p. 1 - 19.

[11] G. Shimura, Introduction to the arithmetic theory of automorphic functions, Iwanami Shoten Publishers and Princeton Univ. Press, Publications of the Math. Soc. of Japan vol. 11, 1971.

ERRATUM

Matrices M = f U, U a unimodular matrix yield the unit correspondence which is excluded from T(n)'.

Chapter IV: THE BASIS PROBLEM

§1 Introduction

In the following we restrict ourselves to modular forms with principal character $\chi(G) = 1$. The elements $\theta(z)$ in a suitable linear combination of the columns of the matrix series II, (16) span a C-Module $\Theta_l(D, H)$. We are interested in the case $l = k-2$. For $k = 2$ we replace the $B_0(n) = B_0(n; D, H)$ by their components $B_0'(n) = B_0'(n; D, H)$, explained in corollary 1 to theorem 2 in II. The elements in one column form the module $\Theta_0'(D, H)$. With this all elements of $\Theta_{k-2}(D, H)$ and $\Theta_0'(D, H)$ are cusp forms. Lastly we denote the C-Module of all $\theta(K z)$ for $\theta(z) \in \Theta_{k-2}(D, H)$ resp. $\Theta_0'(D, H)$ with $\Theta_{k-2}(D, H)^K$ resp. $\Theta_0'(D, H)^K$.

Proposition *Let* χ_1 *be the principal character mod* DH, *defined as* $\chi_1(a) = 1$ *or* 0, *according as* $(a, DH) = 1$ *or not. Then*

$$(1)\quad \theta(Kz; D, H)\, T(n) = B_{k-2}(n; D, H, \chi_1)\, \theta(Kz; D, H), \text{ for } (n,K) = 1.$$

The same holds in the case $k = 2$ *for the* θ_0' *and* B_0'.

Proof. At first let p be a prime which does not divide H. The definition of T(p) in III, (17) shows

$$\theta(Kz)\, T(p) = \sum_{n=1}^{\infty} \chi_1(p)p^{k-1}\, B_{k-2}(n)e^{2\pi iKpnz} + \sum_{n=1}^{\infty} B_{k-2}(n)e^{2\pi iKp^{-1}nz}$$

(in the second term all summands with non-integral exponent $p^{-1}n$ are omitted). This is

$$= \sum_{n=1}^{\infty} \left(\chi_1(p)p^{k-1}B_{k-2}(\tfrac{n}{p}) + B_{k-2}(pn)\right)e^{2\pi iKnz},$$

where $B_{k-2}(\frac{n}{p}) = 0$ if p does not divide n. Because of II, (18) and (20) this is (1) if p does not divide H.

Now let $p|H$. According to theorem 2 and corollary 2 in II we can split up

$$B_{k-2}(n; D, H) = \sum_{\nu} B_{k-2}(n, p^{\nu}; D, H).$$

The matrix $B_{k-2}(p, p; D, H)$ occurring among those on the right side is

Footnote:

Remark: The proof of the following theorem does not exclude that the theta series in one particular column are linearly dependent.

equal to $B_{k-2}(p;\ D,\ H,\ \chi_1)$. The corollary states that

$$B_{k-2}(p;\ D,\ H,\ \chi_1)\ B_{k-2}(n,\ p^{\nu};\ D,\ H) = B_{k-2}(pn,\ p^{\nu+1},\ D,\ H),$$

and the summation over all ν yields

$$B_{k-2}(p;\ D,\ H,\ \chi_1)\ B_{k-2}(n;\ D,\ H) = B_{k-2}(pn;\ D,\ H)$$

which is again (1).

For composite n, we get (1) from II, (18).

Theorem. *Let*

$$N = p_1 \cdots p_r \tag{2}$$

be a decomposition of N *into different primes* p_i. *Then the* C-*module* $S_k(\Gamma_0(N))$ *of all cusp forms of even weight* $k > 2$ *and character* 1 *is the direct sum* (3)

$$S_k(\Gamma_0(N)) = \Theta_{k-2}(p_1, p_2 \cdots p_r) \oplus \Theta_{k-2}(p_2, p_3 \cdots p_r) \oplus \Theta_{k-2}(p_2, p_3 \cdots p_r)^{p_1}$$

$$\oplus \cdots \oplus \textstyle\sum_{K|p_1 \cdots p_{r-1}} \Theta_{k-2}(p_r,\ 1)^K \oplus \textstyle\sum_{K|N} S_k(\Gamma)^K.$$

The same holds for $k = 2$, *but then the* $\Theta_{k-2}(D,\ H)$ *are to be replaced by the* $\Theta_0'(D,\ H)$, *and* $S_2(\Gamma) = 0$.

In a first step we compare the traces of the unit operator T(1) on both sides of (3). The trace of T(1) in $S_k(\Gamma_0(N))$ is equal to the rank, and this is given in III, (29). In the spaces $\Theta_{k-2}(p_i, p_{i+1} \cdots p_r)$ the unit operator is equal to $B_{k-2}(1;\ p_i, p_{i+1} \cdots p_r)$ the trace of which can be obtained from II, theorem 5. Now we remark that the first factor in that formula is

$$\frac{\rho^{k-1} - \bar{\rho}^{k-1}}{\rho - \bar{\rho}} = 4\left(\frac{k-1}{4} - \left[\frac{k}{4}\right]\right) \qquad \text{for } \rho = \sqrt{-1}, \tag{4}$$

$$\frac{\rho^{k-1} - \bar{\rho}^{k-1}}{\rho - \bar{\rho}} = 3\left(\frac{k-3}{3} - \left[\frac{k}{3}\right]\right) \qquad \text{for } \rho = \tfrac{1}{2}(-1+\sqrt{-3}).$$

With (4), the traces of the unit operator T(1) on the left of (3) and the sums of the traces on the right can be checked to be equal. We need this also for $n > 1$ which are rational squares, because T(n) has a summand $\sqrt{n}\ T(1)$.

Eichh-66

Secondly we compare the traces $s_k(T(n))$ on the left and the sum of the traces of the $B_{k-2}(n; D, H)$ and of the $s_k(T(n))$ for $\Gamma = \Gamma_0(1)$ on the right, assuming that $(n, N) = 1$. In the case $k = 2$ we have to remember that $b(n) = \sum_{t|n} t$ according to corollary 1 to theorem 2 in II.

If $N = p$ is a prime, we can now easily complete the proof. We only consider the subring generated by the $T(n)$ with $(n, p) = 1$. The representations of it by the modules on both sides of (3) are equivalent, because the $T(n)$ and the $B(n)$ satisfy the same equations (namely II, (18) and (19)) and have the same traces, and the ring is semisimple. Hence a modular form is a sum of the special forms allowed on the right side and another form $\phi(z)$ whose Fourier expansion is $\sum c_n e^{2\pi i p n z}$. Now $\phi(\frac{z}{p})$ is a modular form of level 1. But such $\phi(z)$ occur already.

§2 Proof of the theorem

We proceed by induction on r in (2), without using the preceding proof for $r = 1$. The theorem is trivially true for $r = 0$. For $r > 1$, we put $N = p N'$ and express the statement as

$$(5) \qquad S_k(\Gamma_0(N)) = \Theta_{k-2}(p, N') \oplus S_k(\Gamma_0(N')) \oplus S_k(\Gamma_0(N'))^p$$

where the last module consists of all $\phi(pz)$ with $\phi(z) \in S_k(\Gamma_0(N'))$. (5) is equivalent to the statement that the $T(n)$ yield equivalent representations of the Hecke ring on both sides. But the action of the $T(n)$ on $\Theta_{k-2}(p, N')$ is not known because it is uncertain that the elements in this space as described above are linearly independent. Therefore we replace the action of the $T(n)$ in $\Theta_{k-2}(p, N')$ by the Brandt matrix $B_{k-2}(n; p, N'\chi_1)$ mentioned in the proposition in §1. We shall prove that now we get equivalent representations on both sides, and this implies lastly that the elements in $\Theta_{k-2}(p, N')$ are linearly independent, and that the action of $T(n)$ here is the Brandt matrix.

Lemma 1. Let $\phi(z)$ be a basis of $S_k(\Gamma_0(N'))$, written as a column vector. We have to distinguish between the Hecke operator $T(n)$ on $S_k(\Gamma_0(N))$ and the Hecke operator $T'(n)$ on $S_k(\Gamma_0(N'))$. Let

$$\phi(z)\ T'(n) = R'(n)\ \phi(z)$$

be the matrix representation of the latter. With these notations the matrix representation of $T(n)$ in $S_k(\Gamma_0(N')) \oplus S_k(\Gamma_0(N'))^p$ is

$$\begin{pmatrix}\phi(z)\\ \phi(pz)\end{pmatrix} T(n) = \begin{cases} \begin{pmatrix} R'(n) & 0 \\ 0 & R'(n)\end{pmatrix}\begin{pmatrix}\phi(z)\\ \phi(pz)\end{pmatrix} & \text{for } n \not\equiv 0 \bmod p, \\[2ex] \begin{pmatrix} -p^{k-1}R'(np^{-2}) & -p^{k-1}R'(np^{-1}) \\ R'(np^{-1}) & R'(n)\end{pmatrix}\begin{pmatrix}\phi(z)\\ \phi(pz)\end{pmatrix} & \text{for } n \equiv 0 \bmod p \end{cases}$$

which will be abbreviated by

$$\begin{pmatrix}\phi(z)\\ \phi(pz)\end{pmatrix} T(n) = \tilde{R}(n) \begin{pmatrix}\phi(z)\\ \phi(pz)\end{pmatrix}. \tag{6}$$

Of course $R'(n\,p^{-\nu}) = 0$ if $n\,p^{-\nu} \not\equiv 0 \bmod 1$.

Proof. The case $n \not\equiv 0 \bmod p$ is trivial. For $n \equiv 0 \bmod p$ we remember that $T(n)\ T(p) = T(n\,p)$. The corresponding formula for the $R(n)$ is checked by means of II, (19). Thus we have only to treat the case $n = p$, and then (6) is clear.

For the proof of (5) we have to show that the $T(n)$ yield equivalent representations on both sides. By the proposition in §1 and lemma 1, this amounts to

$$s_k(T(n)) = s(B_{k-2}(n;\ p,\ N',\ \chi_1)) + s(\tilde{R}(n)) \tag{7}$$

with $\tilde{R}(n)$ defined by (6). The trace of $B_{k-2}(n;\dots)$ is given in theorem 7 in II. The other trace is (with obvious abbreviation)

$$s(\tilde{R}(n)) = s_k'(T(n)) - p^{k-1}s_k'(T(np^{-2})) \tag{8}$$

Lemma 2.

$$s(\tilde{R}(n)) = \begin{cases} 2s_k'(T(n)) & \text{if } n \not\equiv 0 \bmod p, \\ s_k'(T(n)) & \text{if } n \equiv 0 \bmod p,\ n \not\equiv 0 \bmod p^2. \end{cases} \tag{9}$$

$s_k'(T(n))$ is the trace given in the theorem of III, with character 1 and

Eich-68

level N' instead of N.

If $n \equiv 0 \bmod p^2$, $s(\tilde{R}(n))$ is obtained from the trace formula in III by the following corrections:

a) in the first line multiply each summand by

$$C_p(n, s, f) = \begin{cases} 1 & \text{for } s \not\equiv 0 \bmod p, \\ 0 & \text{for } s \equiv 0 \bmod p; \end{cases}$$

b) in the second line restrict the summation to such divisors t of n for which $(t, \frac{n}{t}, p) = 1$;

c) the third line replace by

$$n^{\frac{k}{2}-1} (p - 1) (\dim S_k(\Gamma_0(N') + \tfrac{1}{2}) \quad \text{or } 0,$$

d) in the fourth line restrict the summation in the same way as in the second;

e) the fifth line replace by 0.

Proof. Apart from the first, all corrections are immediate consequences of (8).

The first line of the trace formula is the contribution of the substitutions with matrices $M = \begin{pmatrix} A & B \\ C & D \end{pmatrix}$ with fixed points in H (see III, §6). Since $AD - BC = n \equiv 0 \bmod p^2$, there exists a $G \in \Gamma_0(N')$ such that

(10) $$G^{-1} M G = M' \equiv \begin{pmatrix} s & 0 \\ 0 & 0 \end{pmatrix} \quad \text{or} \equiv \begin{pmatrix} 0 & B_1 \\ 0 & 0 \end{pmatrix} \bmod p,$$

according as $A + D = s$ is $\not\equiv 0 \bmod p$ or $\equiv 0 \bmod p$. Each of these normal forms mod p is uniquely determined by M, including the residues B_1 mod p. The M with given n, s, f, and $\Delta = (s^2 - 4n) f^{-2}$ are grouped in $\Pi_{p'/N'} (1+\{\frac{\Delta}{p'}\}) \frac{h(\Delta)}{w(\Delta)}$ classes of equivalent ones, according to proposition 4 in II. The contribution of these M to the trace formula is

(11) $$\frac{\rho^{k-1}-\bar{\rho}^{k-1}}{\rho-\bar{\rho}} \Pi_{p'/N'}(1+\{\frac{\Delta}{p'}\}) \frac{\chi_{p'}(A_{p'})+\chi_{p'}(D_{p'})}{2} \frac{h(\Delta)}{w(\Delta)}.$$

The contribution of such substitutions to $s_k'(T(n\, p^{-2}))$ can be written

Eich-69

$$p^{2-k}\,\frac{(p\rho_1)^{k-1}-(p\bar{\rho}_1)^{k-1}}{p\rho_1-p\bar{\rho}_1}\,\Pi_{p'/N'}\;(1+\{\tfrac{\Delta_1}{p'}\})\;\frac{\chi_{p'}(A_{p'})+\chi_{p'}(D_{p'})}{2}\;\frac{h(\Delta_1)}{w(\Delta_1)}$$

where $\rho_1^2 - s_1\rho_1 + n_1 = 0$ and $n = p^2 n_1$.

It is also obtained by considering the matrices

$$M_1 = \begin{pmatrix} p\,A_1 & B_1 \\ p^2 C_1 & p\,D_1 \end{pmatrix}$$

of trace $s = p\,s_1$ and determinant $n = p^2 n_1$. Each of them occur among the former $M = \begin{pmatrix} A & B \\ C & D \end{pmatrix}$. Each of the residue classes B_1 mod p contribute the same to the trace, and there are p residue classes. Hence the contribution of the M with $s \equiv 0 \bmod p$, $n \not\equiv 0 \bmod p^2$ to the difference (8) is 0, while the contribution of the M with $s \not\equiv 0 \bmod p$ is equal to (11), as asserted in the lemma.

With lemma 2 we can now compare the traces of T(n) in both sides of (5). What has to be shown is (7), and we even prove that this holds for each line of the trace formula of III. What concerns the first line, it also holds for the summands depending on n, s, f. These consist of several factors of which only one differs in the 3 terms in (7). After cancelling all factors which are equal on the 3 terms, we have to check

$$-\tfrac{1}{2}(1+\{\tfrac{\Delta}{p}\})(\chi_{1p}(A_p)+\chi_{1p}(D_p)) = (1-\{\tfrac{\Delta}{p}\})\chi_{1p}(n) - \begin{cases} 2 \\ 1 \\ C_p(n,s,f) \end{cases}$$

according as n is 0, 1, or ≥ 2 times divisible by p; $C_p(n, s, f)$ is given in lemma 2.

To the second line of the trace formula $s(B_{k-2}(\ldots))$ contributes nothing, and the sums $\sum t^{k-2}\,\chi_1(v(n, t, N_1))$ occurring on both sides are equal, due to lemma 2.

In the third line of the trace formula we have

$$n^{\frac{k}{2}-1}\,\chi_1(n)(\dim S_k - \frac{\sqrt{n}-1}{2}\,2^r) = -n^{\frac{k-1}{2}}\,\chi_1(n)\;2^r + n^{\frac{k}{2}-1}\,\chi_1(n)(\dim S_k + 2^{r-1}).$$

The first summand can be combined with the second line by allowing also

$t = \sqrt{n}$ which must be counted with multiplicity $\frac{1}{2}$. The rest is $n^{\frac{k}{2}-1} \chi_1(n)$ times the rank, and we have already proved that the rank on the left of (4) equals the sum of the (formal) ranks on the right.

For the fourth and fifth lines the arguments are the same as for the second and third. With this the proof of the theorem is complete.

§3 Open problems; remarks on the literature

By changing the order of the factors in (2) we obtain linear relations between the theta series for different quaternion orders. This implies that the spaces $\Theta_{k-2}(p_i, p_{i+1} \cdots p_r)$ will often split into subspaces which are invariant under the $T(n)$, especially for $(n, p_i \cdots p_r) = 1$.

Linear dependencies between theta series were first discovered by Witt [9] in a special case. Kitaoka [4] discussed conditions under which two quadratic forms have the same theta series. Kneser [6] pointed out classes of sets of quadratic forms whose theta series depend linearly on each other. He considers quadratic forms which are attached to lattices in the same metric space. His criterion is applicable on the norm forms of left ideals for different orders in a quaternion algebra.

For definite quadratic forms with a square-free discriminant D the class number grows asymptotically with $|D|^{3/2}$. But their theta series are modular forms of level $|D|$ and character $\chi(n) = \mathrm{sign}(n) \left(\frac{D}{n}\right)$, and the number of linearly independent among them grows only with $|D|$. So we meet again linear relations between theta functions.

All these can be written as such between the corresponding Epstein zeta functions. On the other hand, Shimizu [8] gave a number of linear relations between zeta functions of (indefinite) quaternion algebras. His methods consist also of the application of trace formulas.

Our theorem has first been proved by the author [2] in the special case $k = 2$ and $N = p$, a prime. Some steps towards a generalization have been taken in [3]. But no proofs were given, and the trace formula there contains an error. Simultaneously with the preparation of this report the author learned that Hijikata and Saito [10] gave another proof of our theorem and even of a generalization, namely of (5) with $N = p\,N'$, where N' is arbitrary but only $(p, N') = 1$.

The author has shown [2] that modular forms of even weight $k > 2$ and character 1 and of prime level $N = p$ can be written as linear combination of theta series of quadratic forms in $2k$ variables, provided that the field of modular functions of this level is not hyperelliptic. This is indeed the case when N is large. More generally, Lehner and Newman [7] have discussed the conditions under which the field is not hyperelliptic. Also Kitaoka [5] gives spaces of modular forms which are spanned by theta functions.

Lastly we mention that the Hecke operator $T(n)$ maps sums of theta functions on sums of theta functions, if n is the norm of a similarity transformation of the quadratic forms occurring [1, chapter IV]. The proof uses sets of similarity transformations, and these act on the theta series in the same way as the Hecke operators.

Eich-72

References

[1] M. Eichler, Quadratische Formen und orthogonale Gruppen, Springer-Verlag, Berlin-Göttingen-Heidelberg 1952.

[2] ----------, Über die Darstellbarkeit von Thetafunktionen durch Thetareihen, Journ. reine angew. Math. 195 (1956), p. 156 - 171.

[3] ----------, Quadratische Formen und Modulfunktionen, Acta Arithmetica 4 (1958), p. 217 - 239.

[4] Y. Kitaoka, On the relation between positive quadratic forms with the same representation numbers, Proc. Japan Akad. 47 (1971), p. 439 - 441.

[5] ----------, On a space of some theta functions, Nagoya Math. Journ. 42 (1971), p. 89 - 93.

[6] M. Kneser, Lineare Relationen zwischen Darstellungsanzahlen quadratischer Formen, Math. Ann. 168 (1967), p. 31 - 39.

[7] J. Lehner and M. Newman, Weierstrass points of $\Gamma_0(n)$, Annals of Maths. 79 (1964), p. 360 - 368.

[8] H. Shimizu, On zeta functions of quaternion algebras, Annals of Maths. 81 (1965), p. 166 - 193.

[9] E. Witt, Eine Identität zwischen Modulformen zweiten Grades, Abh. Math. Sem. Univ. Hamburg 14 (1941), p. 323 - 337.

[10] H. Hijikata and H. Saito, On the representability of modular forms by theta series, to appear in Number Theory, Algebraic Geometry, and Commutative Algebra, a volume published in honour of Yasuo Akizuki, Tokyo 1973.

APPENDIX: Modular forms with respect to $\Gamma(N)$

We use the following notation:

$$\Gamma_0^0(N) = \{\left(\begin{smallmatrix} a & b \\ c & d \end{smallmatrix}\right) \in \Gamma:\ b \equiv c \equiv 0 \bmod N\}.$$

As is well known, the **C**-module of modular forms (or cusp forms) is the direct sum of modules consisting of forms which behave as

$$f(z)[\begin{smallmatrix} a & b \\ c & d \end{smallmatrix}]^{-k} = \chi(a)\ f(z), \qquad \left(\begin{smallmatrix} a & b \\ c & d \end{smallmatrix}\right) \in \Gamma_0^0(N),$$

with a character $\chi(a)$ mod N^2. The following is obvious:

If $f(z)$ *is such a form,* $f(Nz)$ *is a form with respect to* $\Gamma_0(N^2)$ *with the same character, and conversely.*

Modular forms of different characters are linked by an operator which has first been used by Weil [2] in a special case, and for quite a different purpose. In the explanation we use the abbreviation $[a, b, \ldots]$ for the smallest common multiple of natural integers $a, b, \ldots$.

Theorem 1. *Let* $f(z)$ *be a modular form with respect to* $\Gamma_0(N)$ *and character* χ *which is defined mod* M*, a divisor of* N*. Furthermore let* ψ *be a character mod* L*. Then*

$$f(z)\ \Phi(\psi) = \sum_{\lambda \bmod L} \psi(\lambda)\ f(z + \tfrac{\lambda}{L}) = g(z)$$

is a modular form (of the same weight) with respect to $\Gamma_0([N, ML, L^2])$ *and character*

$$\chi' = \chi\ \psi^2$$

which is defined mod $[M, L]$.

For an n *prime to* L*, the Hecke operators* $T_\chi(n)$ *and* $T_{\chi'}(n)$ *on* $f(z)$ *and* $g(z)$ *commute in the following way:*

$$T_\chi(n)\Phi(\psi) = \psi(n)\Phi(\psi)\ T_{\chi'}(n).$$

Proof. We may write more briefly

Eich-74

$$\Phi(\psi) = \sum_\lambda \psi(\lambda)\left[\begin{smallmatrix}1 & \lambda L^{-1}\\ 0 & 1\end{smallmatrix}\right]^{-k}$$

and find for $\left(\begin{smallmatrix}a & b\\ c & d\end{smallmatrix}\right)$, $\left(\begin{smallmatrix}a' & b'\\ c' & d'\end{smallmatrix}\right) \in \Gamma$:

$$\Phi(\psi)\left[\begin{smallmatrix}a & b\\ c & d\end{smallmatrix}\right]^{-k} = \sum_\lambda \psi(\lambda)\left[\begin{smallmatrix}a+c\lambda L^{-1} & b+d\lambda L^{-1}\\ c & d\end{smallmatrix}\right]^{-k},$$

$$\left[\begin{smallmatrix}a' & b'\\ c' & d'\end{smallmatrix}\right]^{-k} \Phi(\psi) = \sum_{\lambda'} \psi(\lambda')\left[\begin{smallmatrix}a' & b+a'\lambda'L^{-1}\\ c & d'+c\lambda'L^{-1}\end{smallmatrix}\right]^{-k}.$$

We assume

$$c = c' \equiv 0 \mod [N,\ ML,\ L^2]$$

which implies

$$a\,d \equiv a'd' \equiv 1 \mod [N,\ ML,\ L^2].$$

The matrices occurring on the right can be replaced by others, for instance the latter by $\left(\begin{smallmatrix}a'' & b''\\ c'' & d''\end{smallmatrix}\right)$ with

$$a'' \equiv a' \mod M, \qquad b'' \equiv b + a'\lambda'L^{-1} \mod 1,$$

$$c'' = c, \qquad d'' \equiv d' + c\lambda'L^{-1} \mod M.$$

We put

$$\left.\begin{aligned} a'' &= a + c\lambda L^{-1}\\ d'' &= d = d' + c\lambda'L^{-1}\end{aligned}\right\} \text{ which is } \equiv \begin{cases} a' & \mod [M,\ L],\\ d' + c\lambda'L^{-1} & \mod [M,\ L],\end{cases}$$

and for the second summation index we take

$$\lambda' = \frac{d}{a}\lambda \mod [M,\ L] \text{ which is } \equiv \frac{d}{a'}\lambda \mod [M,\ L],$$

and which implies

$$b + d\lambda L^{-1} \equiv b + a'\lambda'L^{-1} \mod 1.$$

With this we find

$$\Phi(\psi)\left[\begin{smallmatrix}a & b\\ c & d\end{smallmatrix}\right]^{-k} = \psi(a)^2 \left[\begin{smallmatrix}a' & b\\ c & d'\end{smallmatrix}\right]^{-k} \Phi(\psi)$$

and

$$f(z)\Phi(\psi)\left[\begin{smallmatrix}a & b\\ c & d\end{smallmatrix}\right]^{-k} = \chi(a)\psi(a)^2\, f(z)\Phi(\psi).$$

Lastly we apply $T_\chi(n)$ to $f(z)$ with $(n,\ L) = 1$:

$$f(z)T_\chi(n) = n^{\frac{k}{2}-1} \sum_{a,b,d} \bar{\chi}(a)\, f(\tfrac{az+b}{d})\, (\tfrac{a}{d})^{\frac{k}{2}}$$

where a, b, d run over

$$a\,d = n,\ a > 0,\ b \bmod d.$$

Applying $\Phi(\psi)$ we get

$$f(z)\, T_\chi(n)\Phi(\psi) = n^{\frac{k}{2}-1} \sum_\lambda \psi(\lambda) \sum_{a,b,d} \bar{\chi}(a)\, f(\tfrac{az+b}{d} + \tfrac{a}{d}\tfrac{\lambda}{L})$$

where $\frac{a}{d}\lambda$ can be replaced by an integral λ' in the same congruence class mod L. Then the right side becomes

$$= n^{\frac{k}{2}-1} \sum_{a,b,d} \bar{\chi}(a)\bar{\psi}(\tfrac{a}{d}) \big(f(z')\Phi(\psi)\big)\Big|_{z' = (az+b)d^{-1}}$$

$$= \psi(n)\, f(z)\Phi(\psi)\, T_{\chi'}(n), \qquad \text{q. e. d.}$$

We conclude with some remarks on the Hecke ring for modular forms with respect to the principal congruence subgroup $\Gamma(N)$. The matrices

$$A_t \in \Gamma:\ \ A_t \equiv \begin{pmatrix} t & 0 \\ 0 & t^{-1} \end{pmatrix} \bmod N$$

for a t which is prime to N are automorphisms of the field of modular functions for $\Gamma(N)$. They commute with the Hecke operators with $(n, N) = 1$.

(1) $$T(n) = \sum_{a,b,d} [A_t^{-1}]\, \begin{bmatrix} a & bN \\ 0 & d \end{bmatrix}$$

(which have to carry a superscript -k when applied to forms of weight k). Under the Hecke ring we understand the ring generated by the T(n) and A_t.

Now we assume that the congruence

(2) $$n \equiv t^2 \bmod N$$

be solvable, and with a solution we put

(3) $$\hat{T}(n) = [A_t]\, T(n).$$

Eich-76

From (1) - (3) it is obvious that, for all $G \in \Gamma$:

$$\hat{T}(n)\, G = G\, \hat{T}(n). \tag{4}$$

We are interested in the subring S of the Hecke ring, generated by these $\hat{T}(n)$.

Let R be a C-module of integral modular forms with respect to $\Gamma(n)$ which is invariant under Γ and under the T(n). Then R can be decomposed into a direct sum

$$R = R_1 \oplus R_2 \oplus \ldots \tag{5}$$

where the R_i yield irreducible representations of the modular congruence group

$$M(N) = \Gamma/\Gamma(N).$$

Any of the irreducible representation modules P_j ($j = 1, \ldots, h$) of $M(N)$ occurs in (5) in a certain multiplicity t_j, i.e. t_j of the R_i are equivalent with P_j. Let r_j be the degree of P_j.

Theorem 2. R *can also be decomposed into a direct sum*

$$R = S_1 \oplus S_2 \oplus \ldots \tag{6}$$

where the S_j *are representation modules of the ring* S *of degrees* t_j, *and every* S_j *occurs (up to equivalence) in (6) exactly* r_j *times.*

Proof. A $G \in \Gamma$ is represented by a matrix consisting of blocks along the diagonal of r_j lines, and each block appears t_j times, if a suitable basis according to (5) is underlying. Subdivide a representation of $\hat{T}(n)$ into blocks of corresponding sizes. Then (4) and the irreducibility of R_1 implies that the first t_j blocks are of the form

$$\begin{pmatrix} E_{r_1} t_{11} & \cdots & E_{r_1} t_{1t_1} \\ \cdots & \cdots & \cdots \\ E_{r_1} t_{t_1 1} & \cdots & E_{r_1} t_{t_1 t_1} \end{pmatrix} = E_{r_1} \times (t_{ij})$$

with the unit matrix E_{r_1} of r_1 rows, while in the other places of the first $r_1 t_1$ lines and columns stands 0.

Simultaneously we obtain the following formulas for the traces $s(G)$, $s(T(n))$ of the representations of a $G \in M(N)$ and a $\hat{T}(n)$ in R, and the traces $s_j(G)$ and $s_j(\hat{T}(n))$ of G in P_j and $\hat{T}(n)$ in S_j:

(7) $$s(G) = \sum_{j=1}^{h} t_j s_j(G),$$

(8) $$s(\hat{T}(n)) = \sum_{j=1}^{h} r_j s_j(\hat{T}(n)),$$

(9) $$s(\hat{T}(n)G) = \sum_{j=1}^{h} s_j(\hat{T}(n)\, s_j(G).$$

The $s_j(\hat{T}(n))$ have been determined by the author [1] under the following assumptions:

a) The representation P_j is already uniquely determined by the square of the trace $s_j(G)$, which is often the case.

b) the weight k is even.

c) R is the space of all cusp forms.

For the trace $s(\hat{T}(n))$ the assumption a) is superfluous.

The assumption b) could be dropped if the methods of this report (III) are adopted. Surprisingly the considerations for a general N are not more complicated than for a square-free N. But the formula for $s_j(\hat{T}(n))$ contains the traces $s_j(G)$ which are in general not yet known.

[1] M. Eichler, Einige Anwendungen der Spurformel im Bereich der Modularkorrespondenzen, Math. Annalen 168 (1967), 128 - 137.

[2] A. Weil, Über die Bestimmung Dirichletscher Reihen durch Funktionalgleichungen, Math. Annalen 168 (1967), 149 - 156.

(Note : For most of the errata I thank Henri Cohen and Don Zagier, especially those on ps.133, 134. They concern the final trace formula for T(n) with n a rational square. The correction on p.138 is due to Paul Ponomarev and Arnold Pizer. Ponomarev found an example that the theta series in one column of the matrices $\theta(n)$ can be linearly dependent.

Cohen and Zagier pointed out that, on p.134, we need not restrict ourselves to even weights, since irregular cusps do not exist if N is square-free. Furthermore, the formula (29) may be written more generally

$$\dim S_k(\Gamma_0(N),\chi) = \frac{k-1}{12}\prod_{p|N}(p+1) - \frac{1}{2}\prod_{p|N} 2$$

$$-(\frac{k-1}{4} - [\frac{k}{4}])\chi(a_2)\prod_{p|N}(1 + (\frac{-4}{p}))$$

$$-(\frac{k-1}{3} - [\frac{k}{3}])\frac{\chi(a_3) + \bar{\chi}(a_3)}{2}\prod_{p|N}(1 + (\frac{-3}{p})).$$

$$+ \delta(\chi) = \begin{cases} 1 & \text{if } k = 2 \text{ and } \chi \text{ is the principal character} \\ 0 & \text{otherwise} \end{cases}$$

M. Eichler).

CLASS-NUMBERS OF COMPLEX QUADRATIC FIELDS

By H. M. Stark

International Summer School on Modular Functions

Antwerp 1972

CONTENTS

1. Complex multiplication

Let E be an elliptic curve

$$y^2 = 4x^3 - g_2x - g_3, \qquad \Delta = g_2^3 - 27g_3^2 \neq 0,$$

in Weierstrass normal form. The curve may be parametrized by the Weierstrass $\wp$-function, $x = \wp(z)$, $y = \wp'(z)$. The function $\wp(z)$ is a doubly periodic function whose periods form a lattice

$$\Lambda = \{\omega_1,\omega_2\} = \{a\omega_1 + b\omega_2 \mid a,b \in \mathbb{Z}\}$$

where $\omega_1/\omega_2 \notin \mathbb{R}$ and for convenience we assume that ω_1 and ω_2 are so ordered that $\mathrm{Im}(\omega_1/\omega_2) > 0$. We then have the relations

$$(1) \qquad g_2 = 60 \sum_{\omega}{}' \omega^{-4}, \qquad g_3 = 140 \sum_{\omega}{}' \omega^{-6},$$

where the summations are over all $\omega \in \Lambda$ other than $\omega = 0$, and

$$(2) \qquad \wp(z) = \frac{1}{z^2} + \frac{g_2}{20} z^2 + \frac{g_3}{28} z^4 + \frac{g_2^2}{1200} z^6 + \dots \quad .$$

If $f(z + \omega) = f(z)$ for all ω in Λ, we say f is a function on $\mathbb{C}/\Lambda$. Any meromorphic function $f(z)$ on $\mathbb{C}/\Lambda$ is of the form $r(\wp) + \wp' s(\wp)$ where r and s are rational functions of $\wp(z)$ and if $f(z)$ is an even function then $f(z) = r(\wp(z))$ only. In particular if $\beta \in \mathbb{C}$ is such that $\wp(\beta z)$ is a function on $\mathbb{C}/\Lambda$, then $\wp(\beta z)$ is a rational function of $\wp(z)$. For example, this is true for $\beta = n \in \mathbb{Z}$ where it is well known that $\wp(nz)$ is the quotient of a polynomial in $\wp$ of degree n^2 by a polynomial in $\wp$ of degree $n^2 - 1$ (we call this multiplication on E by n). If for some $\beta \in \mathbb{C} - \mathbb{Z}$, $\wp(\beta z)$ is a function on $\mathbb{C}/\Lambda$ then we say that E admits (complex) multiplication by β. This will happen if and only if there

St-4

are integers r, s, t, u such that

$$(3) \qquad \beta\binom{\omega_1}{\omega_2} = B\binom{\omega_1}{\omega_2}, \qquad B = \begin{pmatrix} r & s \\ t & u \end{pmatrix}$$

(this is of course also true for $\beta \in \mathbb{Z}$). In particular β is an eigenvalue of B and so is an algebraic integer which must be of degree two since we are assuming $\beta \notin \mathbb{Z}$. Further, $\beta = t(\omega_1/\omega_2) + u$ and since $\beta \notin \mathbb{Z}$, $t \neq 0$ and hence β and ω_1/ω_2 are in the same quadratic field which must be complex since ω_1/ω_2 is not real. Let the quadratic field be $\mathbb{Q}(\sqrt{d})$ where $d < 0$ is the discriminant of the field. If E admits multiplication by β_1 and β_2 then we see from (3) (or directly) that E admits multiplication by $\beta_1 \pm \beta_2$ and $\beta_1\beta_2$. Thus the set of all α such that E admits multiplication by β is a subring of the integers of $\mathbb{Q}(\sqrt{d})$ containing 1 and is thus, if not just $\mathbb{Z}$, an order in $\mathbb{Q}(\sqrt{d})$.

We are interested in the case when E admits multiplication by the whole ring of integers o of $\mathbb{Q}(\sqrt{d})$. Let

$$\delta = \begin{cases} \dfrac{1 + \sqrt{d}}{2} & d \equiv 1(4) \\[2ex] \frac{1}{2}\sqrt{d} & d \equiv 0(4) \end{cases}$$

so that 1 and δ are an integral basis for o. Thus E admits multiplication by o if and only if there are integers r, s, t, u such that (3) holds with $\beta = \delta$ and this happens if and only if ω_1/ω_2 is in $\mathbb{Q}(\sqrt{d})$ and $\{1,\omega_1/\omega_2\}$ is a fractional ideal of $\mathbb{Q}(\sqrt{d})$ with integral basis 1, ω_1/ω_2. Conversely, if 1, ω_1/ω_2 form an integral basis for a fractional ideal of $\mathbb{Q}(\sqrt{d})$ then E admits complex multiplication by o.

Let

$$j(E) = j(\Lambda) = j(\omega_1/\omega_2) = 1728g_2^3/\Delta.$$

The j-function is an analytic function of ω_1/ω_2 (in the upper half plane) and depends only on Λ or E. Two curves E_1 and E_2 are birationally equivalent if and only if they have the same j-invariant and this happens if and only if there exists a constant ζ such that the corresponding lattices satisfy $\Lambda_2 = \zeta\Lambda_1$. If Λ_2 and Λ_1 are fractional ideals of $\mathbb{Q}(\sqrt{d})$ the relation $\Lambda_2 = \zeta\Lambda_1$ implies $\zeta \in \mathbb{Q}(\sqrt{d})$ and Λ_2 and Λ_1 are in the same ideal class and conversely. Thus if h(d) is the class-number of $\mathbb{Q}(\sqrt{d})$, then there are (up to birational equivalence) exactly h(d) curves with complex multiplication by $\mathfrak{o}$.

The question arises as to how to find the result of complex multiplication by β when it exists. Suppose E admits complex multiplication by β. Then

$$\wp(\beta z) = \frac{f(\wp(z))}{g(\wp(z))} \tag{4}$$

where f and g are polynomials of degree $N(\beta)$ and $N(\beta) - 1$, respectively (the degrees of f and g are determined by examining the poles of $\wp(\beta z)$). We find f and g by continued fractions. For real numbers α, the continued fraction algorithm is given by

$$\alpha_o = \alpha; \quad a_j = [\alpha_j], \quad \alpha_j = a_j + \alpha_{j+1}^{-1},$$

where α_{j+1} is defined provided $\alpha_j \neq [\alpha_j]$. We set

$$\frac{p_j}{q_j} = a_o + \cfrac{1}{a_1 + \cfrac{1}{a_2 + \cfrac{\ddots}{\quad \cfrac{1}{a_{j-1} + \cfrac{1}{a_j}}}}}$$

St-6

If α is rational then $\alpha_n = a_n$ for some n and $\alpha = \frac{p_n}{q_n}$. Further we find that if we set $q_{-2} = 1$, $p_{-2} = 0$, $q_{-1} = 0$, $p_{-1} = 1$ then for $j \geq 0$,

$$(q_j, p_j) = (q_{j-2}, p_{j-2}) + a_j(q_{j-1}, p_{j-1}).$$

For example if $\alpha = \frac{5}{3}$, we have

	j	a_j	q_j	p_j
	-2		1	0
	-1		0	1
5/3 = 1 + 2/3,	0	1	1	1
3/2 = 1 + 1/2,	1	1	1	2
2/1 = 2	2	2	3	5

We see in general that p_n and q_n are relatively prime since the above recursion may be rewritten as

$$\begin{matrix} q_{j-1} & p_{j-1} \\ q_j & p_j \end{matrix} = \begin{matrix} 0 & 1 \\ 1 & a_j \end{matrix} \begin{matrix} q_{j-2} & p_{j-2} \\ q_{j-1} & p_{j-1} \end{matrix}$$

and so $q_{n-1}p_n - p_{n-1}q_n = (-1)^{n-1}$.

The same algorithm works for rational functions. For example, if $\alpha = -\frac{2x^2 + 4x + 9}{4x + 8}$ then

	j	a_j	q_j	p_j
	-2		1	0
	-1		0	1
$-\frac{2x^2 + 4x + 9}{4x + 8} = -\frac{x}{2} - \frac{9}{4x + 8}$,	0	$-\frac{x}{2}$	1	$-\frac{x}{2}$
$-\frac{4x + 8}{9} = -\frac{4x + 8}{9}$,	1	$-\frac{4x + 8}{9}$	$-\frac{4x + 8}{9}$	$\frac{2x^2 + 4x + 9}{9}$.

Here again p_n and q_n are relatively prime polynomials.

In the problem above, we want to find polynomials f and g such that (4) holds. We do this by continued fractions. For example, suppose we are given that

$$E: \quad y^2 = 4x^3 - 30x - 28 \qquad (j = 8000)$$

admits complex multiplication by $\sqrt{-2}$. How do we find $\wp(z\sqrt{-2})$? We have from (2),

$$\wp(z) = \frac{1}{z^2} + \frac{3}{2}z^2 + z^4 + \dots$$

$$= \frac{1}{z^2}\,(1,0,0,0,\tfrac{3}{2},0,1,\dots)$$

where by $(c_o,c_1,\dots)$ we mean $c_o + c_1 z + \dots$. Thus

$$\alpha_o = \alpha = \wp(z\sqrt{-2}) = -\frac{1}{2z^2}\,(1,0,0,0,6,0,-8\dots)$$

$$= -\frac{1}{2}\wp(z) + \frac{1}{2z^2}\,(0,0,0,0,-\tfrac{9}{2},0,9,\dots)$$

$$= -\frac{1}{2}\wp(z) - \frac{9}{4}z^2(1,0,-2,\dots).$$

Now $a_o = -\frac{1}{2}\wp(z)$ and

$$\alpha_1 = (\alpha_o - a_o)^{-1} = -\frac{4}{9z^2}\,(1,0,2,\dots)$$

$$= -\frac{4}{9}\wp(z) - \frac{8}{9} + \frac{1}{z^2}\,(0,0,0,\dots),$$

and thus $a_1 = -\frac{4}{9}\wp(z) - \frac{8}{9}$. This is the example above for rational functions with $x = \wp(z)$ and so

St-8

$$\frac{p_1}{q_1} = -\frac{2\wp(z)^2 + 4\wp(z) + 9}{4\wp(z) + 8} .$$

Since $N(\sqrt{-2}) = 2$ and since the p_n and q_n are relatively prime and increasing in degree, we must in fact have $\alpha_1 = a_1$ and $\wp(z\sqrt{-2}) = p_1/q_1$.

But how do we know that this curve admits complex multiplication by $\sqrt{-2}$? Since $h(-8) = 1$ there is just one class of curves admitting complex multiplication by $\sqrt{-2}$ and the period lattice for such a curve is given by $\Lambda = \zeta\{1,\sqrt{-2}\}$ for some ζ. By (1), we may choose ζ so that g_2/g_3 has any desired value (the exceptions are $\omega_1/\omega_2 = i$ when $g_3 = 0$, $j = 1728$, and $\omega_1/\omega_2 = \rho$, $\rho^3 = 1$, when $g_2 = 0$, $j = 0$). Ordinarily, $g_2 = g_3$ would be suitable, but here $g_2/g_3 = 30/28$ is numerically better. Thus for some g, $g_2 = 30g$ and $g_3 = 28g$ $(g \neq 0)$. We now use one more term in the series expansion of $\wp(z)$:

$$\wp(z) = \frac{1}{z^2}\,(1,0,0,0,\tfrac{3}{2}g,0,\tfrac{3}{4}g^2,\ldots),$$

$$\wp(z\sqrt{-2}) = -\frac{1}{2z^2}\,(1,0,0,0,6g,0,-8,0,12g^2,\ldots).$$

Proceeding as above, we get

$$a_o = -\tfrac{1}{2}\wp(z), \qquad a_1 = \frac{4}{9g}\wp(z) - \frac{8}{9g}$$

and

$$\alpha_1 - a_1 = \frac{16}{9g}\, z^2(g-1,\ldots).$$

To have complex multiplication by $\sqrt{-2}$, we must have $\alpha_1 - a_1 = 0$ and hence $g = 1$, $j(\sqrt{-2}) = 8000$.

Let us now return to a general curve E which admits complex multiplicatio by o. We assume $d < -4$ since we know $j(\sqrt{-1}) = 1728$ and $j(\rho) = 0$ already. We may scale Λ so that $g_2 = g_3$ (= g say) without changing j(E). Now

$$\wp(z) = \frac{1}{z^2}\,(1,0,0,0,\frac{g}{20},0,\frac{g}{28},\ldots)$$

$$\wp(\delta z) = \frac{1}{\delta^2 z^2}\,(1,0,0,0,\frac{g}{20}\delta^4,0,\frac{g}{28}\delta^6,\ldots)$$

We go through the continued fraction algorithm which will terminate with $\alpha_n = a_n$ when we get to a q_n with degree $N(\delta) - 1$ in $\wp(z)$. Since not all curves possess complex multiplication by δ, the series for $\alpha_n - a_n$ expressed in terms of g and z has an infinite number of powers of z which have non-zero polynomials in $\mathbb{Q}(\sqrt{d})(g)$ as coefficients. Thus the value of g corresponding to E is algebraic and thus so is

$$j = \frac{1728g}{g - 27} .$$

Let σ be an automorphism of the algebraic numbers over $\mathbb{Q}$ and let E^σ be the curve

$$E^\sigma:\ y^2 = 4x^3 - g^\sigma x - g^\sigma .$$

By conjugating every step of the continued fraction algorithm, we see that E^σ admits complex multiplication by δ^σ and hence by $o = \{1,\delta\} = \{1,\bar{\delta}\}$. There exist h(d) different classes of curves admitting multiplication by o and thus j is an algebraic number of degree $\leq h(d)$. In particular if $h(d) = 1$ then $j(\delta) \in \mathbb{Q}$.

In actual fact, if $\tau \in \mathbb{Q}(\sqrt{d})$, $\operatorname{Im} \tau > 0$, $j(\tau)$ is an algebraic integer and if $\{1,\tau\}$ is a fractional ideal of $\mathbb{Q}(\sqrt{d})$ then $j(\tau)$ has degree h(d) over $\mathbb{Q}(\sqrt{d})$ (and $\mathbb{Q}$ also) and $\mathbb{Q}(\sqrt{d})(j(\tau))$ is the absolute class field of $\mathbb{Q}(\sqrt{d})$ (see for example, [8, §13] or the lectures of Shimura at this sum-

St-10

2. Complex quadratic fields with class-number one

Suppose $Q(\sqrt{d})$ has $h(d) = 1$ and $d < -8$. If d is even, 2 ramifies and so is a norm of some integer, $2 = x^2 + \frac{|d|}{4} y^2$ which is impossible. Thus d is odd. Again if $d \equiv 1 \pmod 8$, 2 splits into the product of two primes and so 2 is a norm, $2 = x^2 + xy + \frac{|d| + 1}{4} y^2$ which is impossible. Therefore $d \equiv 5 \pmod 8$, $|d| \equiv 3 \pmod 8$. Further $|d|$ is a prime since otherwise (with $d \neq -4$ or -8) $h(d)$ would be even. In like manner if $d \equiv 1 \pmod 3$ then 3 is a norm and so $d < -11$ implies $d \equiv 2 \pmod 3$, $|d| \equiv 1 \pmod 3$ and hence $|d| \equiv 19 \pmod{24}$. Throughout this section, we shall assume $|d| \equiv 3 \pmod 8$, $3 \nmid d$.

The Fourier series for $j(z)$ is

$$j(z) = \frac{1}{q} + 744 + 196884q + 21493760q^2 + \ldots$$

where as usual, $q = e^{2\pi i z}$. For example, if $z = (1 + \sqrt{-163})/2$ then

$$q^{-1} = -e^{\pi\sqrt{163}} = -262\ 537\ 412\ 640\ 768\ 743.999\ 999\ 999\ 999\ 2\ldots$$

and so, since $h(-163) = 1$, we may say with complete assurance (that means, without bothering to analyze the error terms) that

$$j\left(\frac{1 + \sqrt{-163}}{2}\right) = -262\ 537\ 412\ 640\ 768\ 000 .$$

This number is $(-640320)^3$, a perfect cube!

Let

$$\gamma_2(z) = [j(z)]^{1/3},$$

the cube root being chosen which is real on the imaginary axis. Since the only zeros of $j(z)$ in the upper half plane, $\mathfrak{H}$, are triple zeros, $\gamma_2(z)$ is analytic in $\mathfrak{H}$. Since $j(z + 1) = j(z)$, we see that

$$\gamma_2(z + 1) = e^{-\frac{2\pi i}{3}} \gamma_2(z),$$

the correct cube root being easily determined when Im z is large. Let δ have the same meaning as in the last section. When $3 \nmid d$, the real cube root of $j(\delta)$ is also of degree $h(d)$ (this is also true for $d = -3$). We denote this real cube root by γ:

$$\gamma = e^{-\frac{2\pi i}{3}} \gamma_2(\delta)$$

Let

$$f(z) = e^{-\frac{2\pi i}{48}} \frac{\eta(\frac{z+1}{2})}{\eta(z)} = q^{-1/48} \prod_{n=1}^{\infty} (1 + q^{n-\frac{1}{2}}).$$

The function $f(2z - 1)$ is a root of

$$x^{24} + e^{-2\pi i/3} \gamma_2(z)x^{16} - 256 = 0.$$

In particular $f = f(\sqrt{d}) = f(2z-1)$ is a root of

$$f^{24} + \gamma f^{16} - 256 = 0 .$$

Thus f^8 is an algebraic integer of degree $\leq 3h(d)$.

Weber [9] proved that when $|d| \equiv 3 \pmod 8$, $3 \nmid d$ that f^8 is of degree exactly $3h(d)$. He further proved f^2 is of degree $3h(d)$ and conjectured that the same is true of f itself. All this is true for $d = -3$ also since then $\gamma = 0$. This conjecture was first proved four years ago by Birch [1]. For the rest of this section, we presume $h(d) = 1$, $d < -8$.

Then $\mathbb{Q}(f) = \mathbb{Q}(f^2) = \mathbb{Q}(f^4) = \mathbb{Q}(f^8)$ is a cubic extension of $\mathbb{Q}$. Thus there is a unique monic cubic equation for f^k ($k = 1,2,4,8$) over $\mathbb{Q}$. Say this equation is given by f being a root of

$$x^{3k} + B_k x^{2k} + A_k x^k + C_k = 0$$

where A_k, B_k, C_k are in $\mathbb{Q}$ and in fact $\mathbb{Z}$. If we transpose $B_k x^{2k} + C_k$ to the other side, square and transpose back, we find for $k = 1,2,4$,

$$(7) \qquad 2A_k - B_k^2 = B_{2k}, \quad A_k^2 - 2B_k C_k = A_{2k}, \quad -C_k^2 = C_{2k} .$$

But $B_8 = \gamma$, $A_8 = 0$, $C_8 = -256$. In particular $C_4 = \pm 16$ and in fact -16 since $C_4 = -C_2^2$. In like manner $C_2 = -4$. From $A_8 = 0$, we now find

$$A_4^2 + 32B_4 = 0$$

and if we insert $A_4 = A_2^2 + 8B_2$, $B_4 = 2A_2 - B_2^2$ into this and rearrange the terms, we get

$$A_2^4 - 64A_2 = 2(A_2^2 + 4B_2)^2.$$

Hence $2 \mid A_2$. It now follows that $4 \mid A_2$ and $2 \mid B_2$. Set $A_2 = -4\alpha$, $B_2 = -2\beta$. Then

$$(8) \qquad 2\alpha(\alpha^3 + 1) = (2\alpha^2 - \beta)^2 .$$

It follows that α is of the form $\pm x^2$ or $\alpha = \pm 2x^2$.

LEMMA 1: <u>The only solutions</u> $x,y,z \in \mathbb{Z}$, $(x,y) = 1$, <u>to</u>

i) $x^6 + y^6 = 2z^2$;

ii) $x^6 - y^6 = 2z^2$;

iii) $8x^6 + y^6 = z^2$;

iv) $8x^6 - y^6 = z^2$

<u>are given by</u>

i) $x^2 = y^2 = 1$;

ii) $x^2 = y^2 = 1$;

iii) $x^2 = y^2 = 1$ <u>and</u> $x = 0$, $y^2 = 1$;

iv) <u>none at all</u>; <u>respectively</u>.

The proof for part iv) is easy since the equation (mod 8) shows y is even and thus $8 \mid z^2$, $4 \mid z$, and so $2 \mid x$ and $(x,y) \neq 1$ (and in fact the only solution is $x = y = z = 0$). The other three cases are considerably more difficult but can be handled by decent arguments. In any event, our immediate application comes with $y = 1$ in which case the proof of cases i), ii), iii), is decidedly easier. Cases i, ii, iii, iv of the Lemma arise from (8) (with $y = 1$) when $\alpha = x^2$, $\alpha = -x^2$, $\alpha = 2x^2$, $\alpha = -2x^2$ respectively. Thus the only solutions to (8) are given by

$$(\alpha,\beta) = (0,0),\ (1,0),\ (-1,2),\ (2,2),\ (1,4),\ (2,14).$$

This leads back to six values of γ,

$$\gamma = 0,\ -32,\ -96,\ -960,\ -5280,\ -640320$$

respectively. We already know that $d = -3, -11, -19, -43, -67, -163$ satisfy all the necessary conditions and so these six values of γ must correspond to these discriminants (and in this order since $j(z)$ is real valued and monotonic on the line $\mathrm{Re}\, z = 1/2$). In particular each solution corresponds to a known field with $h(d) = 1$ and thus there are no other $d < -8$ with $h(d) = 1$. Together with $d = -4, -7, -8$, we have thus proved

St-14

THEOREM 1: <u>If</u> $h(d) = 1$, $d < 0$, <u>then</u> $d = -3, -4, -7, -8, -11, -19, -43, -67, -163$.

The method of proof of Theorem 1 given here is essentially due to Heegner [4]. His proof was, however, unclear and the reader may wish to consult the commentaries in [1], [2], [3], [5], [6]. One of the remarkable things about this proof is that the Diophantine equation (8) has no extraneous solutions. Because of Lemma 1, we have found all rational solutions to (8) with $\alpha = \pm(x/y)^2$ and $\alpha = \pm 2(x,y)^2$. However, there are other points on the elliptic curve (8) as $\alpha = 1/23$ demonstrates.

For the class-number one problem however, we have more information. Let $A_1 = -2\tau$, $B_1 = -2\rho$ (we take $C_1 = -2$ since otherwise we could replace f by -f in everything to date). Then

$$2\rho - \tau^2 = \alpha, \qquad 2(\rho^2 + \tau) = \beta.$$

We substitute this into (8) and seek all rational solutions. It is convenient to throw out some of the information in this equation by letting

$$w = 1/\tau^3, \qquad z = \frac{2\rho-\tau^2}{\tau^2} = \alpha/\tau^2.$$

We may then group all the terms so as to get

$$\tfrac{1}{8}z(z^4 + 4z^3 - 2z^2 + 4z + 1) = [(z-2)w + \tfrac{1}{2}(3z^2 - 2z - 1)]^2.$$

This equation is similar to (8) but with one important difference: the degree of the left side of (9) is odd. If we write $z = \frac{p}{q}$ with $q > 0$, $(p,q) = 1$ then it is a simple matter to show from (9) that q is either a square or twice a square and up to sign the same is true of p. Hence $z = \pm\square$ or $\pm 2.\square$ and therefore the same is true of α ($\tau = 0$ gives $\alpha = 0$

or 2 directly). By Lemma 1 we have still only the six solutions above for (α,β). Thus there are not even any extraneous rational solutions for ρ and τ.

3. Some units related to all this

In this section it will be assumed throughout that $|d| \equiv 7 \pmod{12}$. If $3\sqrt{3} \mid (\gamma + 6)$ then at $z = \delta$ the roots' $\xi(z)$ of

$$(10) \qquad \xi(z) - \xi(z)^{-1} = \frac{-1}{3\sqrt{3}} (e^{-\frac{2\pi i}{3}} \gamma_2(z) + 6)$$

are units and conversely. Solving this equation for $\xi = \xi(\delta)$ gives

$$\xi = \frac{-(\gamma+6)\sqrt{3} \pm [\frac{j-1728}{\gamma-12} \cdot 3]^{1/2}}{18}$$

where $j = \gamma^3 = j(\delta)$. The purpose of expressing the solution this way is that by Weber [9] $j-1728 = \gamma_3(\delta)^2$ is d times a square in $\mathbb{Q}(j)$ and by Heegner [4] (proved by Birch [2]), $\gamma-12$ is -3 times a square in $\mathbb{Q}(j)$. Thus ξ is of the form $\mu\sqrt{3} + \nu\sqrt{|d|}$ with μ and ν in $\mathbb{Q}(j)$ and ξ^2 is in $\mathbb{Q}(\sqrt{d},j)(\sqrt{-3})$. We take for $\xi(z)$ the root which tends to ∞ as Im $z \to \infty$. Thus $\xi(\delta)$ is the positive root of (10) when $z = \delta$.

A more familiar way to obtain units is from the Δ-function. Let

$$\varepsilon(z) = 3^{-12} \frac{\Delta(\frac{z}{3})\Delta(\frac{z+2}{3})}{\Delta(3z)\Delta(\frac{z+1}{3})}$$

Then $\varepsilon(z) + \varepsilon(z)^{-1}$ is a function invariant under S^3: $z \to z + 3$, T: $z \to -1/z$, $S^{-1}TS$ and STS^{-1}. These generate a group G_3 which also leaves $\gamma_2(z)$ invariant and $\gamma_2(z)$ is a function of order one on $\mathfrak{H}/G_3$ (a fundamental domain $\mathcal{D}$ for G_3 is the usual fundamental domain $\mathcal{D}_o$ for the modular group together with translates of $\mathcal{D}_o$ by ± 1. This was mentioned in Birch's lectures when he discussed cycloidal groups). Thus since $\gamma_2(z)$ has a pole (of order 1) in $\mathcal{D}$ only at i_∞ and $\varepsilon(z) + \varepsilon(z)^{-1}$ has a pole (of order 8) in $\mathcal{D}$ only at i_∞, it follows that $\varepsilon(z) + \varepsilon(z)^{-1}$ is an eighth degree polynomial in $\gamma_2(z)$.

THEOREM 2: $\xi^8 = \varepsilon(\delta)$.

One method of proof is to set

$$\xi_1(z) = \frac{\zeta_8}{3\sqrt{3}} \frac{\eta(\frac{z}{3})^3 \ \eta(\frac{z+2}{3})^3}{\eta(3z)^3 \ \eta(\frac{z+1}{3})^3}$$

where $\zeta_n = e^{2\pi i/n}$ and ζ_8 is the eighth root of unity which makes $\xi_1(z)$ real and positive for Re $z = \frac{1}{2}$. Set

$$g(z) = \xi_1(z) - \xi_1(z)^{-1} + \frac{1}{3\sqrt{3}} (e^{-\frac{2\pi i}{3}} \gamma_2(z) + 6).$$

From

$$\eta(z)^3 = q^{1/8} (1 - 3q + 5q^3 + O(q^5))$$

we find that

$$(11) \quad \xi_1(z) = -\frac{1}{3\sqrt{3}} \zeta_3^{-1} q^{-1/3}(1+6\zeta_3 q^{1/3}+27\zeta_3^2 q^{2/3}+86q+243\zeta_3 q^{4/3}+O(q^{5/3}))$$

and thus $g(z) = O(q^{4/3})$ and is 0 at i_∞. If we have sufficient energy we can now verify that either g_1 is zero at all rational cusps or that g is invariant under S^3, T, $S^{-1}TS$ and STS^{-1} and hence identically zero in either case. This would involve keeping track of which eighth root of unity arises in transformations of η^3. However, this is not necessary.

To find $\xi(z)^8 + \xi(z)^{-8}$, we square (10), add 2 to both sides, square, subtract 2 and square again. Since

$$\xi_1(z) - \xi_1(z)^{-1} = \xi(z) - \xi(z)^{-1} + O(q^{4/3})$$

we see that

$$\xi_1(z)^8 - \xi_1(z)^{-8} = \xi(z)^8 - \xi(z)^{-8} + O(q^{-1}).$$

In other words

$$[\varepsilon(z) + \varepsilon(z)^{-1}] - [\xi(z)^8 + \xi(z)^{-8}] = h(z)$$

where h(z) is a third degree polynomial in $\gamma_2(z)$. If we had carried out (11) to $O(q^{-9/3})$ we would have had enough information necessary to find $h(z) = O(q^{1/3})$ and hence $h(z) = 0$ for all z. But this also is not necessary. We need only verify the result of the Theorem four times and h(z) will then be identically zero.

4. Kronecker's limit formula

Again in this section, $|d| \equiv 7 \pmod{12}$. The question arises as to how we may verify Theorem 2 numerically. For this purpose, we recall Kronecker's limit formula. If

$$Q(x,y) = ax^2 + bxy + cy^2$$

$$= a(x+\theta y)(x+\bar{\theta}y)$$

where $a > 0$, $b^2 - 4ac < 0$, $\operatorname{Im}\theta > 0$ then

St-18

$$\lim_{s\to 1^+}\left[\frac{1}{2}\sum_{x,y\neq 0,0}\sum Q(x,y)^{-s} - \frac{1}{|d|^{s/2}}\left(\frac{\pi}{s-1} + \text{const.}\right)\right]$$

$$= \frac{\pi}{12\sqrt{|d|}}\log\{(\theta - \bar{\theta})^{12}\,|\Delta(\theta)|^2\}.$$

We may generalize this to allow characters as coefficients. Let

$$L(s,\chi_k,Q) = \frac{1}{2}\sum_{x,y\neq 0,0}\sum \chi_k(Q(x,y))Q(x,y)^{-s}$$

where $\chi_k(n) = (\frac{k}{n})$ is the Kronecker symbol defined for discriminants, k, of quadratic fields and for k = -3, $(\frac{-3}{n}) = (\frac{n}{3})$ where the latter is the Legendre symbol. We see that if $\mathfrak{a}$ is an integral ideal of $\mathbb{Q}(\sqrt{d})$, then $\psi(\mathfrak{a}) = \chi_{-3}(N\mathfrak{a})$ is a ring class character (mod 3) and with $Q_1(x,y) = x^2 + xy + \frac{|d| + 1}{4} y^2$,

$$L(s,\chi_{-3},Q_1) = \sum_{\mathfrak{a}} \psi(\mathfrak{a})N\mathfrak{a}^{-s}.$$

There are, for $|d| \equiv 7 \pmod{12}$, exactly 4 ring classes and the sum over each ring class may be related back to Kronecker's limit formula. The result is [7],

$$L(1,\chi_{-3},Q_1) = \frac{\pi}{36\sqrt{|d|}}\log\left|\frac{1}{3^{12}}\,\frac{\Delta(\frac{\delta}{3})\Delta(\frac{\delta+2}{3})}{\Delta(3\delta)\Delta(\frac{\delta+1}{3})}\right|^2$$

(12)

$$= \frac{\pi}{18\sqrt{|d|}}\log(\varepsilon(\delta)).$$

When h(d) = 1, we have

$$L(s,\chi_{-3},Q_1) = L(s,\chi_{-3})L(s,\chi_{-3d})$$

where

$$L(s,\chi) = \sum_{n=1}^{\infty} \chi(n)n^{-s}$$

is an ordinary Dirichlet L-series. We may thus evaluate $L(1,\chi_{-3},Q_1)$ from Dirichlet's class-number formula. We then find for $h(d) = 1$ (and $|d| \equiv 7 \pmod{12}$) that

$$\varepsilon(\delta) = \varepsilon_{3|d|}^{4h(3|d|)}$$

where for $k > 0$, ε_k is the fundamental unit of $\mathbb{Q}(\sqrt{k})$, and thus

$$\xi_1(\delta) = \varepsilon_{3|d|}^{h(3|d|)/2}$$

For $d = -7, -19, -43, -67$, $h(3|d|) = 1$ and

$$\varepsilon_{21}^{1/2} = \frac{\sqrt{3}+\sqrt{7}}{2}, \qquad \varepsilon_{57}^{1/2} = 5\sqrt{3} + 2\sqrt{19},$$

$$\varepsilon_{129}^{1/2} = 53\sqrt{3} + 14\sqrt{43}, \qquad \varepsilon_{201}^{1/2} = 293\sqrt{3} + 62\sqrt{67}.$$

Writing these as $\mu\sqrt{3} + \nu\sqrt{|d|}$, we have $3\mu^2 - |d|\nu^2 = -1$ (congruences (mod 3) demonstrate that = 1 cannot occur) in each case and $-18\mu - 6 = -15, -96, -960, -5280$ respectively.

This provides us with the four needed verifications of the Theorem (the last three values have been shown to be γ by Heegner's method also) and completes the proof. We may check the result for $d = -163$ also. The quickest way to find ε_{489} is by continued fractions. We have

$$\sqrt{489} = [22,\overline{8,1,4,1,1,1,3,2,1,2,14,2,1,2,3,1,1,1,4,1,8,44}]$$

where (see (5) and (6) for the notation) $a_o = 22$, $a_1 = 8,\ldots,a_{22} = 44$

St-20

and the bar means that $a_{j+22} = a_j$ for $j \geq 1$. It is well known that $p_{21} + q_{21}\sqrt{489}$ is either ε_{489} or ε_{489}^3 and in fact since $489 \equiv 1(8)$,

$$\varepsilon_{489} = p_{21} + q_{21}\sqrt{489}.$$

(On the other hand, expanding $\sqrt{21}$ in a continued fraction gives ε_{21}^3; this arises with $d = -7$). However, it is not necessary to go through 21 recursions and then take the square root of ε_{489} since $\varepsilon_{489}^{1/2}$ essentially occurs in the middle of the expansion. In this case

$$\sqrt{3}\ \varepsilon_{489}^{1/2} = p_{10} + q_{10}\sqrt{489}, \qquad \varepsilon_{489}^{1/2} = \frac{p_{10}}{3}\sqrt{3} + q_{10}\sqrt{163}.$$

Thus, since $h(489) = 1$,

$$\gamma = -18(\frac{p_{10}}{3}) - 6 = -6(p_{10} + 1)$$

which checks since $p_{10} = 106719$. Conversely, we may use Theorem 2 to claim that $h(489) = 1$.

We may use Theorem 2 to find γ in some cases when $h(d) \neq 1$. In fact, $L(s,\chi_{-3},Q_1)$ is a combination of Dirichlet L-functions whenever $\mathbb{Q}(\sqrt{d})$ has one class per genus. If we write $d = gt$ where g and t are discriminants of quadratic fields (except $t = 1$, $g = d$ is allowed) with $t > 0$, $g < 0$ in all possible ways (there are 2^{r-1} ways of doing this where r is the number of distinct prime factors of $|d|$) and there is one class per genus ($h(d) = 2^{r-1}$) then

$$(13) \qquad L(1,\chi_{-3},Q_1) = \frac{2\pi}{2^{r-1}.3\sqrt{|d|}} \log\Big(\prod_{g,t} \varepsilon_{-3g}^{h(-3g)h(-3t)\omega_{-3t}^{-1}}\Big)$$

where $\omega_{-3t} = \begin{cases} 3 \text{ if } t = 1 \\ 1 \text{ if } t > 1 \end{cases}$. From this and (12) we find that

$$\varepsilon(\delta) = \prod_{g,t} \varepsilon_{-3g}^{3\omega_{-3t}^{-1}\cdot h(-3g)h(-3t)/2^{r-3}}$$

and thus

$$\varepsilon(\delta) = \prod_{g,t} \varepsilon_{-3g}^{3\omega_{-3t}^{-1}h(-3g)h(-3t)/2^{r}}$$

For example, $d = -91$ has $h(d) = 2$ and from

$$\varepsilon_{21}^{3} = 55 + 12\sqrt{21}, \quad h(21) = 1, \quad \varepsilon_{273}^{1/2} = 11\sqrt{3} + 2\sqrt{91}, \quad h(273) = 2,$$

we find

$$\varepsilon(\delta) = \varepsilon_{21}^{3}\,\varepsilon_{273}^{1/2}$$

$$= (605 + 168\sqrt{13})\sqrt{3} + (110 + \tfrac{396}{13}\sqrt{13})\sqrt{91}.$$

Here we have $Q(j) = Q(\sqrt{13})$ and therefore, $\mu = 605 + 168\sqrt{13}$ which gives

$$\gamma = -18\mu-6 = -10896 - 3024\sqrt{13}.$$

This is by far easier than using the infinite series for γ.

We note that (12) and (13) give an explicit value for $\varepsilon(\delta)$ as a rapidly convergent series when $h(d) = 2$. It is this relation and similar relations that allows an effective solution to be given to the problem of finding all $d < 0$ with $h(d) = 2$ from Baker's transcendence results. Because the solution is effective, one can show after considerable numerical work (not yet in print), that if $h(d) = 2$ then $d \geq -427$. However this computation is unrelated to modular functions. A solution to the class-number two problem by Heegner's method is greatly desired.

REFERENCES

[1] B. J. BIRCH: Diophantine analysis and modular functions, in Algebraic Geometry, Oxford (1969), 35-42.

[2] __________: Weber's class invarients, Mathematika 16 (1969), 283-294.

[3] MAX DEURING: Imaginäre quadratische Zahlkörper mit der Klassenzahl Eins, Invent. Math., 5 (1968), 169-179.

[4] KURT HEEGNER: Diophantische Analysis und Modulfunktionen, Math. Z., 56 (1952), 227-253.

[5] H. M. STARK: On the "gap" in a theorem of Heegner, J. Number Theory 1 (1969), 16-27.

[6] __________: Class-number problems in quadratic fields, in Proc. of the International Congress in Nice 1970, Vol. 1, 511-518.

[7] __________: Values of L-functions at s = 1 (I). L-functions for quadratic forms, Advances in Math. 7 (1971), 301-343.

[8] H. P. F. SWINNERTON-DYER: Applications of algebraic geometry to number theory, in Proc. Symposia in Pure Mathematics vol. 20, Amer. Math. Soc., Providence, 1971, 1-52.

[9] H. WEBER: Lehrbuch der Algebra, vol. 3, 3rd ed., Chelsea, New York (1961) (reprint of the 1908 edition).

SOME CALCULATIONS OF MODULAR RELATIONS

B.J. Birch

International Summer School on Modular Functions

Antwerp 1972

Some calculations of modular relations.

1. I will illustrate Ogg's lectures by giving a few computations of modular relations, in an honest nineteenth-century spirit. A modular relation may, I suppose, be practically any pretty relation between modular functions; but classically the greatest interest has always been in the relation between $j = j(z)$ and $j_N = j(Nz)$ for natural numbers N, and that is the case I will keep to.

The direct relation between j and j_N turns out to be frightful, even for small N : however, $\mathbf{C}(j,j_N)$ is precisely the field of functions on the upper half plane H invariant by $\Gamma_0(N)$, so the problem of finding the relation between j and j_N is practically the same as finding a convenient equation for the curve $H/\Gamma_0(N)$. [Recollect that $\mathbf{C}(j)$ is the field of functions invariant by $\Gamma(1)$, that $\mathbf{C}(j_N)$ is the field of functions invariant by

$$({}^N{}_1)^{-1}\Gamma(1)({}^N{}_1), \quad \text{and that} \quad \Gamma_0(N) = \Gamma(1) \cap ({}^N{}_1)^{-1}\Gamma(1)({}^N{}_1).]$$

One may distinguish two approaches. Ideally, one might wish to use nothing more than the geometry of $H/\Gamma_0(N)$, in the spirit of Klein and, more specifically, of Bd. 3 of Weber's Algebra; in practice, one is driven to use every dirty trick one can lay one's hands on; after the manner of Fricke's Algebra. I will spend half my time in working out some very easy examples ($H/\Gamma_0(2)$, $H/\Gamma_0(3)$, ...) by purely geometric techniques; then I will illustrate less scrupulous methods by working out an equation for $H/\Gamma_0(50)$.

2. We try geometric methods, starting at the very bottom; we use no more than the Riemann mapping theorem. We will find the relations between j,

j_2 and j_3, we will give yet another proof that the Fourier series for $j(z)$ has integer coefficients, and we will show why $j(i)$ turns out to be 1728. As it is inconvenient to type pictures, I will describe them instead.

Recollect that we obtain a Riemann surface for $H/\Gamma(1)$ by taking the fundamental domain $\mathcal{R}_1 : |z| > 1$, $|\mathrm{Re}(z)| < \frac{1}{2}$, and identifying the sides of $\mathcal{R}_1$ by $z \to z + 1$ and $z \leftrightarrow -1/z$, so that there is a branch point of order 2 above i and a branch point of order 3 above $\rho = \frac{1}{2}(-1 + \sqrt{-3})$. This Riemann surface may be completed to a sphere by adding the cusp $i\infty$, and taking $q = e^{2\pi iz}$ as local parameter near $i\infty$. Using Riemann's mapping theorem, $j(z)$ is determined uniquely as the univalent function on $H/\Gamma(1)$ normalised so that $j(\rho) = 0$ and so that near $i\infty$ the Fourier series of $j(z)$ starts with $q^{-1} \ldots$.

A convenient fundamental region for $H/\Gamma_0(2)$ is

$$\mathcal{R}_2 : |\mathrm{Re}\ z| < \tfrac{1}{2}, \quad |z - \tfrac{1}{2}| > \tfrac{1}{2}, \quad |z + \tfrac{1}{2}| > \tfrac{1}{2};$$

the sides of $\mathcal{R}_2$ are identified by $z \mapsto z + 1$ and $z \mapsto z/(1 - 2z)$, so that there is a branch point of order 2 above $\frac{1}{2}(1 + i)$, a cusp of width 1 at $i\infty$ and a cusp of width 2 at 0. By Riemann's theorem, there is a univalent function $t_2(z)$ on $H/\Gamma_0(2)$ with a pole at $i\infty$ and a zero at 0, and we can normalise t_2 so that its Fourier series at $i\infty$ starts with $q^{-1} + \ldots$. The involution $W_2 : z \leftrightarrow -1/2z$ acts on $H/\Gamma_0(2)$; $t_2(-1/2z)$ is another univalent function on $H/\Gamma_0(2)$ with the same zeros and poles as $t_2(z)^{-1}$, so

$$t_2(z)\, t_2(-1/2z) = B^2, \quad \text{with} \quad B = t_2(\tfrac{1}{2} + \tfrac{1}{2}i).$$

The function $j(z)$ is now trivalent on $H/\Gamma_0(2)$, and has a pole at the pole of t_2 and a double pole at the zero of t_2; so $j = \mathrm{cubic}(t_2)/t_2{}^2$.

The leading coefficient of the cubic has to be 1 so that the Fourier series on both sides start with q^{-1}; and the branch point above ρ, at which $j = 0$, has been smoothed out, so

$$j = (t_2 - A)^3/t_2^2 \quad \text{with} \quad A = t_2(\rho).$$

Note that A, B and 0 must be unequal, since they are values of t_2 at different points of $H/\Gamma_0(2)$. But now

$$j_2(z) = j(-1/2z) = \frac{(t_2(-1/2z) - A)^3}{t_2^2\ (-1/2z)} = \frac{(B^2 - At_2)^3}{B^4 t_2};$$

the leading term in the Fourier series is now q^{-2}, so $A^3 = -B^4$. Also, the values of t_2 corresponding to $j = j(i)$ are $t_2(\frac{1}{2} + \frac{1}{2}i)$ and $t_2(i)$ taken twice, so

$$(t - A)^3 - t^2 j(i) = (t - t_2(i))^2 (t - B)$$

identically in t. We have now enough information to calculate A, B; we find

$$A = -256, \quad B = -64, \quad t_2(i) = 512, \quad j(i) = 1728.$$

Hence
$$j(z) = \frac{(t_2 + 256)^3}{t_2^2}, \qquad j_2(z) = \frac{(t_2 + 16)^3}{t_2};$$

eliminating t_2 gives

$$j^3 + j_2^3 - j^2 j_2^2 + f(j, j_2) = 0 \qquad (1)$$

where f is a polynomial with integer coefficients, of lower degree. If

$$j = q^{-1} + \sum_{n=0}^{\infty} c_n q^n,$$

then for $N > 0$ the coefficient of q^{N-5} in (1) is

$$-2c_N + g = 0$$

with $g \in \mathbf{Z}[c_0, c_1, \ldots, c_{N-1}]$; so we deduce by induction that

$$c_N \in \mathbf{Z}[\tfrac{1}{2}].$$

Playing the same game with $H/\Gamma_0(3)$ [for which we may take the fundamental region $\mathcal{R}_3$: $|\text{Re } z| < \frac{1}{2}$, $|z - \frac{1}{3}| > \frac{1}{3}$, $|z + \frac{1}{3}| > \frac{1}{3}$] gives

$$j = (t_3 + 27)(t_3 + 243)^3 t_3^{-3} \quad \text{with} \quad t_3(z)t_3(-1/3z) = 729;$$

we obtain a similar relation between j and j_3, and deduce that $c_N \in \mathbf{Z}[\frac{1}{3}]$.

Hence, $$c_N \in \mathbf{Z}.$$

An advantage of this exceedingly geometric method of obtaining modular equations is that the theory works in considerable generality - it works for non-congruence subgroups, and indeed for some arithmetic groups not commensurable with the modular group, for which the function theory may be unfamiliar. However, explicit calculation becomes unduly complicated rather quickly; the most complicated case I myself have worked with these methods is $H/\Gamma_0(64)$. The advantages and, more to the point, the disadvantages are put rather clearly in Atkin and Swinnerton-Dyer's paper on non-congruence subgroups.

The geometric method is particularly appropriate for the various 'cycloidal' subgroups of $\Gamma(1)$; they are described in Prof. Petersson's lectures. For the record, H/G_2, H/G_3, H/G_5 all have genus zero. On these Riemann surfaces, there are univalent functions γ_3, γ_2, g_5 respectively; and we find

$$j - 1728 = \gamma_3^{\,2}, \quad j = \gamma_2^{\,3}, \quad j = g_5^{\,3}(g_5^{\,2} + 5g_5 + 40).$$

3. In more complicated cases, one has to be less scrupulous, and more methodical. We wish to work out an equation for H/G, that is to say, we wish to describe the field of functions on H/G. To do this, we must first get a reasonable picture of H/G so that we know what we are doing, and we must work out the genus and the cusps. The main problem is then to find a good supply of functions on H/G - once one has this, it is relatively easy to find relations by forming combinations of functions which are holomorphic at all points of H and also at all the cusps.

It is enough to spot modular forms, and then their ratios are functions : in fact, as soon as the genus is at least 3, a good method of finding an equation for H/G is to spot a number of independent differentials, and use them as homogeneous coordinates. To find differentials, there are various methods available:

(i) luck;

(ii) theta functions;

(iii) Eichler's trace formula;

(iv) direct computation of the eigenvalues of the Hecke operators, acting on the 1-dimensional homology of H/G.

Methods (iii) and (iv) are both guaranteed to work for $H/\Gamma_0(N)$, and both have been mechanised (by Hans Frey in his Basel thesis, and by D.J Tingley working in Oxford, respectively); both methods are distinctly laborious, even for a computer, but method (iv) is probably best. When they are available, the first two methods have considerable advantages; note that as soon as we have found a single differential, we may usually generate more by operating on it by the Hecke algebra.

Let us deal with a specific example, where we are lucky.

BJB-8

$H/\Gamma_0(50)$

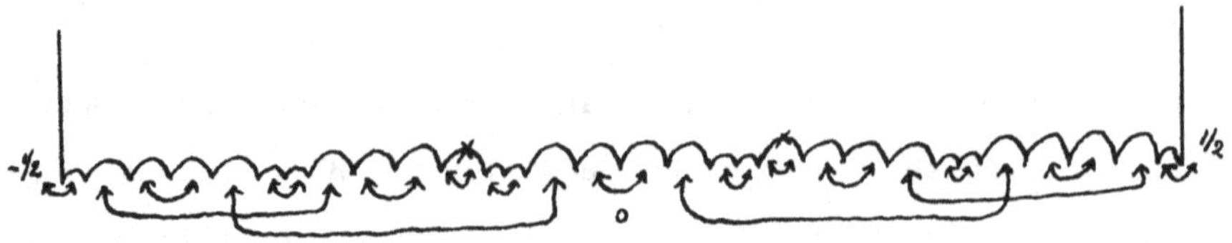

Our picture shows a cusp of width 1 at $i\infty$; a cusp of width 50 at 0; a cusp of width 25 at $\frac{1}{2}$; four more cusps of width 1 at $\pm\frac{1}{10}$ and $\pm\frac{3}{10}$; and five cusps of width 2 at $\frac{1}{25}$, $\pm\frac{1}{5}$ and $\pm\frac{2}{5}$. There are branch points of order 2 at $(i \pm 7)/50$. The index $[\Gamma(1): \Gamma_0(50)] = 90$, and the genus of $H/\Gamma_0(50)$ is 2, consistent with the formula

$$[\Gamma(1) : G] = 12(p - 1) + 6o + 3e_2 + 4e_3.$$

We note that $H/\Gamma_0(50)$ has the usual involution $W_{50} : z \leftrightarrow -1/50z$, and $H/[\Gamma_0(50), W_{50}]$ has genus zero.

This happens to be a case where we can find all the functions we need just by taking products of η-functions. We quote a lemma of Morris Newman [Proc. London Math. Soc. (3) 9 (1959)]; recollect

$$\eta(z) = q^{\frac{1}{24}} \Pi (1 - q^n).$$

LEMMA.
$$\prod_{d|N} \eta(dz)^{r(d)}$$

<u>is a function on</u> $H/\Gamma_0(N)$ <u>so long as</u>

(i) $\sum r(d) = 0$,

(ii) $\Pi\, d^{r(d)}$ <u>is a square</u>,

(iii) $\Pi\, \eta(dz)^{r(d)}$ <u>has integral order at every cusp of</u> $H/\Gamma_0(N)$.

Note that condition (iii) is just a matter of congruences modulo 24.

For example, if N is a prime or a prime square with $(N-1)|24$, then $[\eta(Nz)/\eta(z)]^{24/(N-1)}$ is a function on $H/\Gamma_0(N)$ with a simple zero at $i\infty$ and a simple pole at 0; so the genus of $H/\Gamma_0(N)$ is zero for N = 2,3,4, 5,7,9,13,25.

Returning to $H/\Gamma_0(50)$, the following table gives the valency of $\eta(dz)$ at the various cusps : --

Cusp	$\eta(z)$	$\eta(2z)$	$\eta(5z)$	$\eta(10z)$	$\eta(25z)$	$\eta(50z)$
0	50	25	10	5	2	1
$\frac{1}{2}$	25	50	5	10	1	2
$\pm\frac{1}{5}$, $\pm\frac{2}{5}$	2	1	10	5	2	1
$\pm\frac{1}{10}$, $\pm\frac{3}{10}$	1	2	5	10	1	2
$\frac{1}{25}$	2	1	10	5	50	25
$i\infty$	1	2	5	10	25	50

We deduce that

$$F = \frac{\eta(2z)\eta(25z)}{\eta(z)\eta(50z)}$$

has a simple pole at 0 and $i\infty$ and is finite elsewhere, while

$$G = \frac{\eta^2(2z)\eta(25z)}{\eta(z)\eta^2(50z)}$$

has a 3-fold pole at $i\infty$ and is finite elsewhere. Further, by the functional equation of $\eta(z)$,

$$F|W_{50}(z) = F(-1/50z) = F(z),$$

while $$G|W_{50}(z) = G(-1/50z) = 5F^3/G;$$

so F and $G + 5F^3/G$ are both of them invariant by W_{50}. F is univalent on $H/[\Gamma_0(50), W_{50}]$, so $G + 5F^3/G$ must be a cubic polynomial in F. Now,

$$F = q^{-1} \Pi [(1 + q^n)/(1 + q^{25n})]$$

$$= q^{-1} + 1 + q + 2q^2 + 2q^3 + 3q^4 + 4q^5 + 5q^6 + \dots$$

so $$F^2 = q^{-2} + 2q^{-1} + 3 + 6q + 9q^2 + 14q^3 + 22q^4 + 32q^5 + \dots$$

and $$F^3 = q^{-3} + 3q^{-2} + 6q^{-1} + 13 + 24q + 42q^2 + 73q^3 + 120q^4 + \dots$$

similarly, $$G = q^{-3} + q^{-2} + 1 + q^3 + q^7 + \dots$$

and $$5F^3/G = 5 + 10q + 20q^2 + 40q^3 + 70q^4 + 120q^5 + \dots$$

We deduce $$G + 5F^3/G = F^3 - 2F^2 - 2F + 1.$$

This is an equation for $H/\Gamma_0(50)$; setting $Y = 2G - F^3 + 2F^2 + 2F - 1$ takes it into the convenient hyperelliptic shape

$$Y^2 = (F^3 - 2F^2 - 2F + 1)^2 - 20F^3.$$

What more can be said?

A second involution on $H/\Gamma_0(50)$ is $W_2 : z \leftrightarrow (26z + 1)/(50z + 2)$; we may readily verify that $F|W_2 = -1/F$, $G|W_2 = G/F$, so that $D = G(F + 1)/F^2$ and $E = (F + 1)^2/F$ are invariant by W_2. Then $D^2 - E(E - 5)D + 5E = 0$ is an equation for $H/[\Gamma_0(50), W_2]$, which accordingly has genus 1. Denote $H/[\Gamma_0(50), W_2]$ by ℓ_1 for short; writing $Z_1 = 2D - E(E - 5)$ gives $Z_1^2 = E(E^3 - 10E^2 + 25E - 20)$. To obtain a non-singular plane model for ℓ_1, write $Y_1 = 5Z_1/E^2$, $X_1 = -5/E$; then

$$Y_1^2 = 4X_1^3 + 25X_1^2 + 50X_1 + 25.$$

The elliptic curve ℓ_1 has precisely three rational points, given in this model by the point at infinity and the two points with $X_1 = 0$.

On the other hand, if we take $Z_2 = Z_1(F - 1)/(F + 1)$, then

$$Z_2^2 = (E - 4)(E^3 - 10E^2 + 25E - 20);$$

this is the equation of another elliptic curve ℓ_2 covering $H/\Gamma_0(50)$. Setting $Y_2 = 4Z_2/(E - 4)^2$, $X = -4/(E - 4)$ gives the non-singular model

$$Y_2^2 = 4X_2^3 - 7X_2^2 - 8X_2 + 16;$$

the elliptic curve ℓ_2 has precisely 5 rational points, given in this model by the point at infinity and the four points with $X_2 = 0$ or $X_2 = 2$. The jacobian variety of $H/\Gamma_0(50)$ is isogenous to the product of ℓ_1 and ℓ_2.

The 'twisting operator' R_5 takes a function $f(z)$ to

$$f(z + \tfrac{1}{5}) - f(z + \tfrac{2}{5}) - f(z + \tfrac{3}{5}) + f(z + \tfrac{4}{5});$$

it takes functions on $H/\Gamma_0(50)$ to functions on $H/\Gamma_0(50)$. It gives a 'real multiplication' by $\sqrt{5}$ in the endomorphism ring of the jacobian of $H/\Gamma_0(50)$; in terms of the curves ℓ_1 and ℓ_2, this comes down to the remark that the $\sqrt{5}$ -twist

$$5T^2 = 4X^3 + 25X^2 + 50X + 25$$

of ℓ_1 is isogenous to ℓ_2, and the $\sqrt{5}$ -twist of ℓ_2 is isogenous to ℓ_1. All this is consistent with some computations of Serre (Inventiones Math. 15 (1972); ℓ_2 is the same as Serre's curve 5.7.4.).

Notice that R_5 takes the set of functions whose only pole is at $i\infty$ into itself; in particular, R_5 takes $(G - 1)$ to $-\sqrt{5}(G - 1)$, which explains why most of the coefficients in the Fourier series of G vanish.

ON THE $A_q(\Gamma) \subset B_q(\Gamma)$ CONJECTURE

By Joseph Lehner

International Summer School on Modular Functions

Antwerp 1972

ON THE $A_q(\Gamma) \subset B_q(\Gamma)$ CONJECTURE

1. Let Γ be a Fuchsian group acting on the unit disk $U : |z| < 1$ and let $q \geq 1$; temporarily we assume q integral. The holomorphic function f defined in U is called an automorphic form of <u>weight</u> q (or <u>degree</u> $-2q$) if

(1) $$f(Az)\, A'(z)^q = f(z), \quad z \in U,\ A \in \Gamma.$$

Let R be a normal polygon (fundamental region) for Γ in U. The automorphic form f is called <u>integrable</u>, and we write $f \in A_q(\Gamma)$, if

(2) $$||f||_1 = \int_R\int (1 - |z|^2)^{q-2}\, |f(z)|\,dx\, dy < \infty;$$

it is called <u>bounded</u>, and we write $f \in B_q(\Gamma)$, if

(3) $$||f||_\infty = \sup_{z \in U} (1 - |z|^2)^q\, |f(z)| < \infty.$$

These are the well-known Bers' spaces; see for example [1].
It has been conjectured that

(4) $$A_q(\Gamma) \subset B_q(\Gamma)$$

for all Fuchsian groups Γ. Though this conjecture is still unsettled, it has been verified when Γ is finitely generated. The situation when Γ is of the first kind being essentially trivial, we state the result as follows :

THEOREM. <u>If Γ is a finitely generated group of the second kind, then</u>

$$A_q(\Gamma) \subset B_q(\Gamma).$$

The theorem has been proved (sometimes with additional restrictions) by Drasin and Earle [2], Metzger and Rao [4], [5], and Knopp [3]. A variety of tools has been used in these proofs, in particular, Abel's theorem on Riemann surfaces and a reproducing formula. Here we present a short, elementary proof.

Leh-4

If q is not an integer we assume the existence of a multiplier system

$$\{v(A) : A \in \Gamma;\ |v| = 1\}$$

appropriate to Γ and q; the details are very well known. Then (1) is replaced by

$$f(Az)\, A'(z)^q = v(A)\, f(z),$$

where a definite branch of A'^q has been chosen. In any event

(5) $$|f(Az)|\ |A'(z)|^q = |f(z)|,$$

and this is what we mostly use.

2. Let f be holomorphic in U and set

$$M(\zeta) = (1 - \sigma^2)^q\ |f(\zeta)|, \quad \sigma = |\zeta|,\ \zeta \in U;$$

as is well known $M(A\zeta) = M(\zeta)$ for $A \in \Gamma$. We have

$$f(\zeta) = \frac{1}{\pi a^2} \iint_{\Delta(\zeta)} f(z)\,dx\,dy, \quad z = x + iy,\ a = \frac{1-\sigma}{2}$$

where

$$\Delta(\zeta) = \{z : |z-\zeta| < (1-\sigma)/2\}.$$

Since $1 \leq c_q\ (1 - |z|)^{q-2}/(1 - |\zeta|)^{q-2}$ with

$$c_q = \begin{cases} 2^{q-2}, & q \geq 2 \\ (2/3)^{q-2}, & q < 2 \end{cases}$$

we get

$$M(\zeta) \leq 2^q(1-\sigma)^q\ |f(\zeta)| \leq 2^{q+2}\ \pi^{-1}\ c_q \iint_{\Delta(\zeta)} (1-\rho)^{q-2}\ |f(z)|\,dx\,dy,$$

$$\rho = |z|$$

(6) $$M(\zeta) \leq C \iint_{\Delta(\zeta)} (1-\rho^2)^{q-2}\ |f(z)|\,dx\,dy,\ \zeta \in U$$

where C is a general constant depending only on Γ and q.

Since Γ is of the second kind, the normal polygon R consists of a finite number of sides (circular arcs orthogonal to $Q : |z| = 1$) and a finite (positive) number of free sides (arcs lying on Q). There is exactly one side of R terminating in the left endpoint of a free side and exactly one side terminating in the right endpoint. The quadrilateral bounded by these two sides, the free side, and a circular arc concentric with Q is called a funnel. If Γ has parabolic elements, R will have a finite number of cusps p lying on Q (fixed points of parabolic elements). The region formed by the two sides of R meeting at p and the arc of a circle tangent internally to Q at p is called a cusp sector. R, then, consists of a compact region, a finite number (possibly zero) of cusp sectors, and a finite, positive number of funnels.

A free side of R is bordered in both directions by free sides of other normal polygons. Since all normal polygons have a finite number of sides, there is a number r_0, $0 < r_0 < 1$, with the following property : if ζ lies in a funnel bounded partly by $|z| = r_0$, the disk $\Delta(\zeta)$ intersects not more than two normal polygons. We shall increase r_0, if necessary, so that $|z| = r_0 < 1$ intersects only sides of R that are part of the boundary of a funnel or a cusp sector; r_0 is now fixed and r_0 depends only on Γ.

Let $\zeta \in R$. If $|\zeta| \leq r_0$ we obviously have

$$M(\zeta) \leq C. \tag{7}$$

Next let ζ lie in a funnel, $|\zeta| > r_0$. From (6), the defining property of r_0, and the Γ-invariance of the integrand, we get

$$M(\zeta) \leq 2\ C \int_R\int (1-\rho^2)^{q-2}\ |f(z)|dx\ dy = 2\ C\ ||f||_1, \tag{8}$$

so (7) holds in this case too. Finally, we know that $f(z) \to 0$ as $z \to p$ within a cusp sector at p (cf. §3), so if ζ lies in such a cusp sector, (7) is valid. We have proved (7) when $\zeta \in R$ and it holds when $\zeta \in U$ because of the invariance of $M(\zeta)$ under Γ. Since $||f||_1 = \sup_{\zeta \in U} M(\zeta)$, the theorem is proved.

3. For the convenience of the reader we insert a proof of the well-known fact, mentioned above, that $f(z) \to 0$ as $z \to p$ within a cusp sector at p. Let P generate the stabilizer of p. If $\zeta = Tz = u + iv$ is a conformal map of U on the upper half-plane $H = \{v : v > 0)$ and $T(p) = i\infty$, then $\Gamma_1 = T\,\Gamma\,T^{-1}$ is a discrete group of linear fractional transformations acting on H and $i\infty$ is a cusp of Γ_1, being fixed by $T\,P\,T^{-1} = S$. Of course S is a translation and we may choose T so that S is $\zeta \to \zeta + 1$. S generates the translation subgroup of Γ_1 and $R_1 = T(R)$ is a fundamental region for Γ_1.

Define

$$\hat{f}(\zeta) = f(z)(dz/d\zeta)^q ;$$

it is readily verified that $\hat{f}$ is an automorphic form of weight q on Γ_1 with multipliers

$$v_1(A_1) = v(T^{-1}\,A_1\,T),\ A_1 \in \Gamma_1.$$

Since $\hat{f}(S\zeta) = v_1(S)\,\hat{f}(\zeta)$ and $\hat{f}$ is holomorphic in H, there is a Fourier expansion

$$e^{-2\pi i\kappa\zeta}\,\hat{f}(\zeta) = \sum_{n=-\infty}^{\infty} \hat{a}_n\, e^{2\pi i n\zeta}, \tag{9}$$

where κ is defined by

$$v_1(S) = \exp 2\pi i\kappa, \quad 0 \leqslant \kappa < 1$$

and the series converges uniformly in $v \geqslant v_0 > 0$.

The conformal map T carries the Poincaré metric of U into that of H, i.e.

$$(1 - |z|^2)^{-1} = v^{-1}\,|d\zeta/dz|.$$

A short calculation using (5) then shows that norm is preserved :

$$\text{(10)}\quad ||\hat{f}||_{1,\Gamma_1} = \int_{R_1}\!\!\int v^{q-2}\,|\hat{f}(\zeta)|\,du\,dv = \int_R\!\!\int (1-\rho^2)^{q-2}\,|f(z)|\,dx\,dy = ||f||_1$$

Moreover, R contains a cusp sector ω at p and so R_1 contains a cusp secto[r] $\omega_1 = T\omega$ at $i\infty$; ω_1 is a strip $\{\xi \leqslant u < \xi + 1,\ v \geqslant v_0\}$. Then for $v \geqslant v_0$ we have

$$\hat{a}_n = \int_{\xi}^{\xi+1} \hat{f}(\zeta)\, e^{-2\pi i(n+\kappa)\zeta}\, du,$$

$$|\hat{a}_n| \int_{h_0}^{h} v^{q-2}\, dv \leq C \int_{h_0}^{h} \int_{\xi}^{\xi+1} v^{q-2}\, |\hat{f}(\zeta)| e^{2\pi(n+\kappa)v}\, du\, dv,$$

$$h > h_0 > v_0.$$

It follows that when $m+\kappa \leq 0$,

$$|\hat{a}_n| \leq ||\hat{f}||_{1,\Gamma_1} / C_q(h), \quad C_q(h) = \begin{cases} \dfrac{h^{q-1} - h_0^{q-1}}{q-1}, & q > 1 \\ \log h/h_0, & q = 1 \end{cases}$$

Letting $h \to \infty$ we deduce $\hat{a}_n = 0$ for $n + \kappa \leq 0$, i.e., $n < 0$ when $\kappa > 0$ and $n \leq 0$ when $\kappa = 0$.

Therefore $\hat{f}(\zeta) = O(e^{-Cv})$, $v \to \infty$, or $|f(z)| = |\hat{f}(\zeta)|\ |d\zeta/dz|^{-q} \to 0$ as $z \to p$.

REFERENCES

[1] BERS L., Automorphic forms and Poincaré series for infinitely generated Fuchsian groups, Amer. J. Math. 87 (1965) 196-214.

[2] DRASIN D. and EARLE C.J., On the boundedness of automorphic forms, Proc. Amer. Math. Soc. 19 (1968) 1039-1042.

[3] KNOPP M.I., Bounded and integrable automorphic forms, (to be published).

[4] METZGER T.A. and RAO K.V., On integrable and bounded automorphic forms, Proc. Amer. Math. Soc. 28 (1971) 562-566.

[5] __________________________, On integrable and bounded automorphic forms II., Proc. Amer. Math. Soc. 32 (1972) 201-204.

ADDRESSES OF AUTHORS AND WRITERS

B. BIRCH: Oxford University
Math Institute
24 - 29St Giles
Oxford, England

M. EICHLER: Mathematisches Institut
der Universität Basel
Basel, Switzerland

W. KUYK: Dept. of Mathematics
Rijksuniversitair Centrum Antwerpen
Middelheimlaan 1
2020 Antwerpen, Belgium

J. LEHNER: Dept. of Mathematics
University of Pittsburgh
Pittsburgh, Pennsylvania 15213

A. OGG: Dept. of Mathematics
University of California
Berkeley, California 94720

G. SHIMURA: Dept. of Mathematics
Princeton University
Princeton, New Jersey 08540

H. STARK: Dept. of Mathematics
Massachusetts Institute of Technology
Cambridge, Massachusetts 02139

F. VAN OYSTAEYEN: Dept. of Mathematics
Universitaire Instelling Antwerpen
Fort VI Straat
2610 Wilrijk, Belgium